T0206223

Just Culture

Restoring Trust and Accountability in Your Organization

THIRD EDITION

Just Culture

Restoring Trust and Accountability in Your Organization

THIRD EDITION

SIDNEY DEKKER

CRC Press is an imprint of the
Taylor & Francis Group, an **informa** business

CRC Press
Taylor & Francis Group
6000 Broken Sound Parkway NW, Suite 300
Boca Raton, FL 33487-2742

© 2017 by Taylor & Francis Group, LLC
CRC Press is an imprint of Taylor & Francis Group, an Informa business

No claim to original U.S. Government works

Printed on acid-free paper
Version Date: 20160615

International Standard Book Number-13: 978-1-4724-7578-7 (Paperback)

This book contains information obtained from authentic and highly regarded sources. Reasonable efforts have been made to publish reliable data and information, but the author and publisher cannot assume responsibility for the validity of all materials or the consequences of their use. The authors and publishers have attempted to trace the copyright holders of all material reproduced in this publication and apologize to copyright holders if permission to publish in this form has not been obtained. If any copyright material has not been acknowledged please write and let us know so we may rectify in any future reprint.

Except as permitted under U.S. Copyright Law, no part of this book may be reprinted, reproduced, transmitted, or utilized in any form by any electronic, mechanical, or other means, now known or hereafter invented, including photocopying, microfilming, and recording, or in any information storage or retrieval system, without written permission from the publishers.

For permission to photocopy or use material electronically from this work, please access www.copyright.com (http://www.copyright.com/) or contact the Copyright Clearance Center, Inc. (CCC), 222 Rosewood Drive, Danvers, MA 01923, 978-750-8400. CCC is a not-for-profit organization that provides licenses and registration for a variety of users. For organizations that have been granted a photocopy license by the CCC, a separate system of payment has been arranged.

Trademark Notice: Product or corporate names may be trademarks or registered trademarks, and are used only for identification and explanation without intent to infringe.

Library of Congress Cataloging-in-Publication Data

Names: Dekker, Sidney, author.
Title: Just culture : restoring trust and accountability in your organization / by Sidney Dekker.
Description: Third edition. | Burlington, VT : Ashgate, [2016] | "A just culture is a culture of trust, learning and accountability. It is particularly important when an incident has occurred; when something has gone wrong. How do you respond to the people involved? What do you do to minimize the negative impact, and maximize learning?." | Includes bibliographical references and index.
Identifiers: LCCN 2015043204| ISBN 9781472475756 (hardback : alk. paper) | ISBN 9781472475787 (pbk. : alk. paper)
Subjects: LCSH: Professional ethics.
Classification: LCC BJ1725 .D45 2016 | DDC 174/.4--dc23
LC record available at http://lccn.loc.gov/2015043204

Visit the Taylor & Francis Web site at
http://www.taylorandfrancis.com

and the CRC Press Web site at
http://www.crcpress.com

Contents

Preface

A just culture is a culture of trust, learning, and accountability. A just culture is particularly important when an incident has occurred, when something has gone wrong. How do you respond to the people involved? How do you minimize the negative impact and maximize learning?

The primary purpose of a just culture, to most, is to give people the confidence to report safety issues. Because then people know that the organization will respond fairly. A just culture should enable your organization to learn from an incident, yet also hold people "accountable" for undesirable performance.

RETRIBUTIVE OR RESTORATIVE JUSTICE?

Most of the guidance available on just culture today—and the typical model adopted by many organizations—considers justice in retributive terms. It asks questions such as

- Which rule was broken?
- Who is responsible?
- How bad is the violation (honest mistake, at-risk acts, or reckless behavior) and so what should the consequences be?

Such a "just culture" is organized around shades of retribution. It focuses on the single "offender," asks what they have done and what they deserve. But many managers have found that simplistic guidance about pigeonholing human acts does not take them very far. In fact, it leaves all the hard work of deciding what is just, of what is the right thing to do, to them. And it tends to favor those who already have power in the organization. Simply dividing human behavior up into errors, at-risk acts, or recklessness is really not very helpful. Somebody still needs to decide what category to assign behavior to, and that means that somebody needs to have the power to do so. There is little evidence that organizations with such schemes actually learn more of value from their incidents. Indeed, it is widely held that learning and punishing are mutually exclusive. And should an organization tasked with delivering a product or service be in the business of sanctioning or punishing its people at all?

Fortunately, restorative approaches to justice have been getting more attention. That is a great development—also for organizations wishing to adopt such practices themselves. Restorative justice asks very different questions in the wake of an incident:

- Who is hurt?
- What are their needs?
- Whose obligation is it to meet those needs?

Such an approach to justice and accountability is more inclusive than a retributive one. A variety of people can get hurt by an incident: not just the first victims (patients, passengers) but also the second victim(s): the practitioner(s) involved. Colleagues, the organization, the surrounding community—they too may somehow have been affected by the incident. Hurt creates needs, and needs create obligations. Restorative justices is achieved by systematically considering those needs, and working out collaboratively whose obligation it is to meet them. The second victim may have an obligation to meet the needs of the first, as does the organization. The organization, or colleagues, have obligations toward the second victim (as well as the first). Even first victims can be asked to acknowledge the humanity of the second victim, recognizing that they hurt as well. Reaching a restorative agreement requires that all affected people are involved and have their voices heard. That is hardly ever the case in retributive approaches. Retributive justice might be limited to a boss and an employee. Restorative justice involves them but also the community: first victims of the incident, colleagues, other stakeholders.

In both retributive and restorative approaches, people are held accountable for their actions. Nobody gets "off the hook." Retributive justice asks what a person must do to compensate for his or her action and its consequences: the account is something the person has to *settle*. Restorative justice achieves accountability by listening to multiple accounts and looking ahead at what must be done to repair the trust and relationships that were harmed. This makes it important for others to understand why it made sense for the person to do what they did. Their account is something they *tell*. This also offers an opportunity to express remorse. Restorative approaches are open to multiple voices, and are willing to see practitioners not as offenders or causes of an incident, but as recipients or inheritors of organizational, operational, or design issues that could set up others for failure too. Restorative approaches are therefore more likely to identify the deeper conditions that allowed an incident to happen. They are better at combining accountability and learning, at making them work in each other's favor. Where retributive approaches to a just culture meet hurt with more hurt, restorative approaches meet hurt with healing, and with learning.

JUST CULTURE AND SUSPICIONS ABOUT THE "SYSTEMS APPROACH"

We can trace the popularity of "just culture" to a relatively recent anxiety about "the systems approach." The systems approach says that failure and success are the joint product of many factors—all necessary and only jointly sufficient. When something in your organization succeeds, it is not likely because of one heroic individual. And when something fails, it is not the result of one broken component, or one deficient individual. It takes teamwork, an organization, a system, to succeed. And it takes teamwork, an organization, a system, to fail.

Some people started fearing, though, that this gets people off the hook too easily. That it might allow them to "blame the system" whenever things go wrong. Even when all system provisions to do the right thing were thought to be in place.[1] Safety thinker James Reason asked in 1999: "Are we casting the net too widely in our search for the factors contributing to errors and accidents?"[2] It seemed as if the system approach made it impossible to hold people accountable for poor behavior. Were we no longer allowed to blame anyone? Patient safety champions Bob Wachter and Peter Pronovost observed

> ...beginning a few years ago, some prominent ... leaders began to question the singular embrace of the 'no blame' paradigm. Leape, a patient-safety pioneer and early proponent of systems thinking, described the need for a more aggressive approach to poorly performing [practitioners].[3]

They proposed a retributive just culture program by listing the patient-safety practices that they declared to be beyond the reach of system improvements. Particular attention was given to hand hygiene or signing a surgical site before operating on the patient. Even when everything was in place for clinicians to do the right thing, many still refused, or forgot, or ignored the rules. What should a hospital do with them? Well, said Wachter and Pronovost: hold them accountable. That is, sanction them, punish them. Take a more "aggressive approach." This was no longer a system problem: it was an accountability problem. And so they suggested penalties for individuals who failed to adhere to such practices. These included, for instance, revocation of patient care privileges for certain periods of time. The length of the punishment depended on the severity or frequency of the "offenses" committed by the clinician. Their proposal made it into the prestigious *New England Journal of Medicine*. A swift response by a colleague and me was published there soon as well.[4]

The appeal to punish noncompliant practitioners (or "hold them accountable") is part of a string of such calls during the 1990s and 2000s. Aviation safety researchers, concerned at the enthusiasm of investigators to find all sorts of system precursors and "latent conditions," called at one of their conferences for a refocus on active "errors" by front-line workers rather than mitigating factors in the system.[5] Sometimes, they argued, accidents *are* caused by people at the sharp end—it is as simple as that!

At times, it seemed as if the systems approach itself was on trial.

A concerned surgeon wrote that hospitals have a big problem with unnecessary deaths from medical errors. And the numbers have remained high despite concerted efforts to bring them down. Why? Because we've embraced a so-called solution that doesn't address the problem.

For the last 14 years, the medical profession has put its faith in a systems approach to the problem. The concept was based on the success of such an approach in the field of anesthesia decades ago, but it had been insufficiently tested in medicine overall. And today, despite a widespread embrace of systemized medicine in hospitals across the country, the number of unnecessary deaths hasn't dropped significantly. There's a simple reason for that: Most preventable mishaps in hospitals are caused by the acts of individual practitioners, not flawed systems, and there is plenty of evidence of that fact available.

In 1991, for example, a Harvard Medical Practice Study examined more than 30,000 randomly selected records from 51 hospitals. A table in that study attributed some 61% of harm to patients to either errors of technique (during surgeries and other procedures) or to a failure of doctors to order the correct diagnostic tests.

These are both errors of individuals, not systems. The same study found that only 6% of adverse events were due to systems problems.

And studies have continued to draw similar conclusions. A 2008 analysis of 10,000 surgical patients at the University of South Florida found that, of all the complications among those patients, only 4% were attributable to flawed systems. The rest resulted from individual human shortcomings ... including poor history-taking, inadequate physical examinations, or ordering the wrong tests.[6]

British colleagues agreed, noting the "increasing disquiet at how the importance of individual conduct, performance and responsibility was written out of the ... safety story." To reorient our efforts, they believed, the community would "need to take seriously the performance and behaviors of individual[s]."[7]

My collaborator Nancy Leveson and I wrote spirited responses in the *British Medical Journal*[8] and elsewhere. When did the systems approach become synonymous with blaming the system or its management for problems and shortcomings? The systems approach argues that the interactions produced by the inevitable complexities and dynamics of imperfect systems are responsible for the production of risk—not a few broken components or flawed individuals. It does not eschew individual responsibility for roles taken on inside of those systems. It does not deny accountability that people owe each other within the many crisscrossing relationships that make up such systems. People working in these systems typically don't even want to lose such responsibility or accountability—it gives their work deep meaning and them a strong sense of identity.

I have met "second victims" in a variety of domains.[9] These are practitioners who were involved in an incident for which they feel responsible and guilty. The practitioner might be a nurse, involved in the medication death of an infant, or a surviving pilot who was at the controls during a crash that killed passengers, or an air traffic controller at the radar scope during a collision or near-miss. Never did I get the impression that these people were trying to duck responsibility, that they wanted to avoid being held accountable. In fact, they typically took on *so much* responsibility for what had happened—despite their own ability to point to all the system factors that contributed to the incident—that it led to debilitating trauma-like symptoms. In some cases, this overwhelming sense of personal responsibility and accountability even drove the practitioner to commit suicide.[10]

You will encounter in this book the case of a New Zealand surgeon, who was criminally prosecuted for a number of deaths of patients in his care. What received scant attention was that he was forced to operate with help from medical students, because of a lack of available competent assistance in the hospital that had hired him. Prosecuting the surgeon, who had little control over the context in which he worked, did not solve the problem. It would have similarly affected other clinicians working in that environment. In such cases, blame is the enemy of safety. It finds the culprit and stops any further exploration and conversation. Emphasizing blame and punishment results in hiding errors and eliminates the possibility of learning from them.

Nobody wants to unjustly sanction practitioners for their involvement in an incident. Nobody wants to jeopardize organizational learning by threatening people who disclose their mistakes. Unreflectively or arbitrarily punitive regimes destroy the opportunity to report safety issues without fear of sanction or dismissal. This is why you want to put a "just culture" policy or program in place. It might seem so simple. But justice, accountability, and trust are all hugely difficult to define and agree on. They are what social scientists call "essentially contested" categories. Reasonable, smart people can forever debate their meaning. What is considered just by some might be seen as a deep injustice by others. If you want to get anywhere with a just culture, you have to acknowledge the existence of multiple ways of thinking about those categories. Your point of view is not necessarily right, just like nobody else's is. You have to commit to learning about, valuing, and respecting those other ways too.

A JUST CULTURE HAS MORE ADVANTAGES

The main argument for building a just culture is that *not* having one is bad for both justice and safety. But there is more. Recent research[11,12] has shown that having a just culture can support people's

- Morale
- Commitment to the organization
- Job satisfaction
- Willingness to do that little extra, to step outside their role

Indeed, the idea of justice seems basic to any social relation, basic to what it means to be human. We tend to endow a just culture with benefits that extend beyond making an organization safer. Look at the hope expressed by a policy document from aviation, where a "just culture operates … to foster safe operating practices, and promote the development of internal evaluation programs."[13] It illustrates the great expectations that people endow just cultures with: openness, compliance, fostering safer practices, critical self-evaluation.

Now it may seem obvious why employees may want a just culture. They may want to feel protected from capricious management actions, or from the (as they see it) malicious intentions of a prosecutor. They want to be recognized when they get hurt, they want their needs responded to, and also to have an opportunity to contribute to healing when they can. But this oversimplifies and politicizes things. A just culture, in the long run, benefits everyone.

- **For those who run or regulate organizations, the incentive to have a just culture is very simple. Without it, you won't know what's going on.** A just culture is necessary if you want to monitor the safety of an operation. A just culture is necessary if you want to have any idea about the capability of your people, or regulated organization, to effectively meet the problems that will come their way.
- **For those who work inside an organization, the incentive of having a just culture is not "to get off the hook,"** but to feel empowered to concentrate on doing a quality job rather than on limiting personal liability, to feel

involved and able to contribute to safety improvements by flagging for weak spots, errors, and failures.

- **For those in society who consume the organization's product or service,** just cultures are in their own long-term interest. Without them, organizations and the people in them will focus on better documenting, hiding, or defending decisions—rather than on making better decisions. They will prioritize short-term measures to limit legal or media exposure over long-term investments in safety.

ABOUT THIS BOOK

I wrote the first edition of *Just Culture* (Ashgate, 2007) on the back of a trend toward the criminalization of human error in aviation, healthcare, shipping, and other fields. This concern about criminalization hasn't gone away, to be sure, and the coverage of it still has a prominent place in this third edition. Also, the psychological and sociological mechanisms inherent in criminalization (the power to draw the line, the power to call an act by a particular name, the power to attach sanctions to it) are not very different from what we see in other accountability relationships—even those inside organizations.

But in the years since publication of the first and second editions, I have met managers who struggle with the creation of a just culture themselves. This may or may not be influenced by what happens in society or law or regulations around them. What should they do internally? How should they respond to incidents, errors, and failures that happen on their watch?

So what will you find in the third edition of *Just Culture*? Chapter 1 develops the differences and commonalities between retribution and restoration more fully, giving you a good overview of the options open to you. Chapter 2 then asks a basic but often unasked question: *Why* do your people actually break the rules? It runs through a number of possible explanations, based on what we know from the literature about deviations so far. Each explanation suggests a different managerial or regulatory repertoire of action. Chapter 3 runs through what we currently know about honest disclosure and safety reporting—two of the typical organizational goals that a just culture is supposed to support. Chapter 4 moves out of your organization and into the surrounding environment, focusing on the criminalization of human error. This matters: it sets constraints and creates opportunities for what you can do inside your organization. Chapter 5 summarizes what you might want to do *now*. It puts the things you should probably do and questions you should probably ask in a manageable order and format for you. And it asks, What is the right thing to do when things go wrong? This chapter will take you through some of the basics of moral thinking that might guide you toward ethical answers on the right thing to do.

You will find case studies mixed throughout this book. These sections introduce people's experiences and stories and invite you to reflect on them. Some of the guiding principles are great, you might think. But why are they even necessary? And how would they work in practice? How can you operationalize this? By offering you live experiences from the cases, this book makes the issues more "alive" and helps you address some of those questions.

ABOUT JUST CULTURE, ABOUT ME

A just culture is not a particular program or blueprint. There is no "pure" model that is ideal to implement in any community. You cannot buy a just culture off the shelf from a consultant. Because it won't be a culture, and it will very likely not be just. Nor can you get a just culture by simply reading this book, or by getting your people or managers to read it.

Some of the more exciting and innovative practices for creating trust and accountability in organizational cultures have become visible *after* these things were first written about. They have emerged through dialogue, experimentation, practical innovation, human courage. Of course, the creation of a just culture can be guided by principles. Why would you otherwise even read a book about it? But you should never see these principles—or anybody's principles about just culture—as your algorithm, your policy, your program for how to achieve one. A just culture can only be built from *within* your own organization's practice. The various ideas need to be tried, negotiated, and bargained among the people inside your organization. And you need to test them with the various stakeholders that surround your organization, who want their voices heard when things go wrong. But most of all, you need to test them against your own voice, your own stand.

If you think this is hard, you are right. It is. Creating justice has been one of the most vexing challenges for humanity—ever. You are unlikely to suddenly solve the creation of justice in your own organization in a way that will satisfy everyone. But if you think it is *too* hard, you are wrong. Because not seriously thinking about it will make things even harder. A consultant who promises to deliver you a just culture with some algorithm or program may offer you the illusion that you've solved it. But justice and culture are not for sale. They are things you cannot purchase. Justice and culture emerge from the way you relate to other people, from how you listen to their stories and concerns, from the ethical principles you stand for and from how you govern your business, from the trust you gradually build up: the trust that is so easy to break and so hard to fix. You can be guided in how to do this, for sure, and you can ask or pay others to help you with this. You can be guided to do it better, to do it more humanely. You can be taught to do it with more patience, with more consideration, with more knowledge and wisdom. And I hope that this book will also help you with that.

Since the first edition of *Just Culture* came out in 2007, some people and communities have considered me one of the founding developers and advocates of the concept. This may or may not be true. Over the years, I have in any case tried to remain critical of the various perspectives on just culture (including my own) and open to other ideas. Hence this third edition. It speaks more directly to you if you struggle to create a just culture in your own community. And it examines more broadly and deeply the retributive and restorative options open to you.

But I come with a bias, just like everyone else. I have sat across from second victims on multiple occasions. It moved me to indeed publish a book under that very title (*Second Victim*, CRC Press) in 2013.[9] These encounters inevitably turned the offending practitioners into normal people, hurting people, vulnerable people: people like you and me. These were not people who came to work to do a bad job.

They were not evil or deviant. These were people who did what made sense to them at the time, pretty much like it would have made sense to any of their colleagues. But the consequences of an incident can be devastating, and not just for the first victims. The consequences can overwhelm the most resilient second victims too: even they are often still not adequately prepared—and neither is their organization. In one case, the second victim was murdered by the father and the husband of some of his first victims. In many others, the second victim just muddles through until the end, having lost a job, an identity, a profession, a group of colleagues, a livelihood, a dream, a hope, a life. It is only with considerable support, patience, and understanding that a community around them can be restored, and the trust and accountability that go with it.

So my writing, like any author's, has been formed by those relationships. They have shaped my voice, my vision. In all of this, I endeavor to stay true to my commitments. Even that is an eclectic mix of commitments—a commitment to the voice from below, for instance, and a commitment to justice over power. It involves a commitment to diversity of stories and opinions, but also a commitment to critical thinking, open-mindedness, relentless questioning, and reflection. And it ultimately comes down to a commitment to my own ethical stand: doing our part to help build a world where we respond to suffering not by inflicting even more suffering on each other, but where we respond to suffering with healing.

Acknowledgments

In addition to the many practitioners, students, and colleagues who helped inspire the previous editions of *Just Culture*, I specifically want to acknowledge the contributions of Rob Robson, Rick Strycker, and Hugh Breakey in shaping the ideas for the third edition.

Author

Sidney Dekker (PhD, Ohio State University, USA, 1996) is a professor of humanities and social science at Griffith University in Brisbane, Australia, where he runs the Safety Science Innovation Lab. He is also a professor (hon.) of psychology at the University of Queensland, and professor (hon.) of human factors and patient safety at Children's Health Queensland in Brisbane. Previously, he was a professor of human factors and system safety at Lund University in Sweden. After becoming a full professor, he learned to fly the Boeing 737, working part-time as an airline pilot out of Copenhagen. He has won worldwide acclaim for his groundbreaking work in human factors and safety and is the best-selling author of, among others, *Safety Differently* (2015), *The Field Guide to Understanding 'Human Error'* (2014), *Second Victim* (2013), *Drift into Failure* (2011), *Patient Safety* (2011), and *Ten Questions about Human Error* (2005). More information is available at http://sidneydekker.com/.

Case Study: Under the Gun*

If professionals consider one thing "unjust," it is often this: split-second operational decisions that get evaluated, turned over, examined, picked apart, and analyzed for months—by people who were not there when the decision was taken, and whose daily work does not even involve such decisions.

British special operations officer Christopher Sherwood may just have felt that way in the aftermath of a drug raid that left one man dead.[14] It was midnight, January 15, 1998, when Sherwood and 21 other officers were summoned to the briefing room of the Lewes police station in East Sussex. They had body armor, special helmets, and raid vests (sleeveless vests with two-way radios built in). They may need to immobilize somebody tonight, the briefing began. A raid was mounted, and Sussex was in need of officers who could shoot. Police intelligence had established that a suspected drug dealer from Liverpool and his associates were in a block of flats in St. Leonards, near Hastings. They were believed to be trying to work their way into the extensive drug trade on the British south coast, and to have a kilogram of cocaine with them.

One of the men was James Ashley, previously convicted of manslaughter. The other was Thomas McCrudden, thought to have stabbed a man before. In the briefing, both men were described as violent and dangerous. And most likely armed. The purpose of the raid was to capture the two men and to confiscate their contraband.

As usual with intelligence, however, it was incomplete. Where in the block of flats they were going to be was not known. There were no plans over the flats either—all would have to be searched, and as quickly as possible: with speed and surprise.

Equipped with rifles (fitted with flashlights on the barrel) and automatic pistols, and up to 60 rounds of ammunition each, the officers proceeded in convoy to the Hastings buildings. None of them were in any doubt about the threat awaiting them, nor about the uncertainty of the outcome. "You get out on the plot, and you never, never know how it's going to end," one veteran explained later. "Your heart is pounding...."

After quietly unloading, and making their way to the block in the dark, six officers took up positions outside the target building. The rest were divided up into pairs, accompanied by an officer with an "enforcer," capable of busting front doors and other obstacles. Each group was assigned a specific flat to search, where one officer would cover the left side of whatever room they entered, and the other the right side.

"You know you are walking into danger," commented another officer later. "You know you may be in a situation where you have to kill or be killed. It's a hell of a responsibility."

Christopher Sherwood was one of the officers who went to Flat 6. He was 30 years old, and for carrying that "hell of a responsibility," he was getting paid £20,000 per year (about $35,000 dollars per year at that time).

---

* This case study is taken from Ref. 14.

The door went down under the impact of the enforcer and Sherwood veered into his half of the room. Peering through his gun sight into the dark, he could make out a man running toward him, one arm outstretched. The officer's time was running out quickly. Less than a second to decide what to do—the figure in the dark did not respond, did not stop. Less than two feet to go. This had to be Ashley. Or McCrudden. And armed. Violent, dangerous. Capable of killing. And now probably desperate.

Sherwood fired. Even if there had been time for the thought (which there almost certainly was not), Sherwood would rather be alive and accountable than dead. Most, if not all, officers would. The bullet ripped into the gloaming assailant, knocking him backward off his feet. Sherwood immediately bent down, found the wound, tried to staunch it, searched for the weapon. Where was it?

Screaming started. The lights came on. A woman appeared out of the bedroom and found Sherwood bent over a man flat on the ground—Ashley. Ashley, splayed on his back and bleeding, was stark naked. He was unarmed. And soon dead, very soon. It was determined later that Sherwood's bullet had entered Ashley's body at the shoulder but deflected off the collarbone and gone straight into the heart, exiting through the ribcage. Ashley had died instantly.

Whenever a police officer fired a fatal shot, an investigation was started automatically. It did in this case. The Kent police force was appointed to investigate. They found systemic failure in the Sussex force, including concocted intelligence, bad planning, misapplication of raid techniques, and a wrong focus on small-time crooks. Kent accused Sussex of a "complete corporate failure" in researching, planning, and executing the raid.

Sherwood, devastated that he had killed an unarmed man, was interviewed for four days. He maintained that he, given the knowledge available to him at the time, had acted in self-defense. Not long thereafter, however, Sherwood read that the investigator had prepared reports for the Crown Prosecution Service and the Director of Public Prosecutions, though it was added that nobody knew at that point whether criminal charges were going to be brought.

They were. A year and a half after his shot in the dark, Sherwood was charged with murder.

"Why should anyone want to risk their career or their liberty," an ex-firearms officer reflected, "if, when something goes wrong, they are treated like a criminal themselves?"

The "complete corporate failure" that had sent Sherwood into the building, together with other officers loaded with 1200 rounds of ammunition, faced consequences too (such as admonition letters). Nothing as serious as murder charges.

A review of the armed raid by a superintendent from the National Firearms School had said that there had been no need for firearms to be used. Another superintendent, with responsibility for the guidelines for the use of firearms, said that this case had not met the requirements. The tactic for searching the flats, known as "Bermuda," was also extremely risky. "Bermuda" was originally designed for rescuing hostages from imminent execution. Sussex Police claimed that their inspiration for using "Bermuda" for arresting suspects came from the Royal Ulster Constabulary (RUC) in Northern Ireland. The RUC denied. Sussex Police's own memos had warned as early as 1992 that "risk factors are high and, as such, it should only be considered as a last resort." Their specialist tactical adviser had been warned by the head of the police's National Firearms School that "Bermuda" was too dangerous for such circumstances.

Meanwhile, the inquiry discovered that there had been meetings that had been kept quiet between senior officers and some of those involved in the shooting. After those discoveries, the Kent inquiry stopped cooperation with the chief constable of Sussex, and informed the Police Complaints Authority that they suspected a cover-up. Sussex countered that Kent was bullying and incompetent. The Hampshire police force then took over the investigation instead. Yet in this defensive finger-pointing aftermath, nothing stood out as much as Sherwood's murder charge.

The murder charge may not have been connected only to this case. The Police Complaints Authority was facing a national outcry about the apparent impunity with which officers could get away with shooting people. In the previous 10 years, police in England and Wales had shot 41 unarmed people, killing 15 of them. No police officer had ever been convicted of a criminal offense, and most involved were not even prosecuted. In this case, it seemed as if Sherwood was to be a sacrificial lamb, the scapegoat, so that it would be obvious that the police was finally doing something about the problem. After a year and a half of mulling over a split-second decision, people who had not been there to share the tense, menacing moments in the dangerous dark opined that Chris should have made a different decision. Not necessarily because of Sherwood, but because of all the other officers and previous incidents, the public image of the police service, and the pressure this put on its superiors.

It would not have been the first time that a single individual was made to carry the moral and explanatory load of a system failure. It would not have been the first time for charges against that individual to be about protecting much larger interests. Many cases in this book point in similar directions. What it raises may seem troubling. These sacrifices violate Aristotelian principles of justice that have underpinned Western society for millennia. Such justice particularly means refraining from *pleonexia*, that is, from gaining some advantage for oneself by blaming another, by denying another person what is due to her or him, by not fulfilling a duty or a promise, or not showing respect, or by destroying somebody else's freedom or reputation. Holding back from *pleonexia* puts an enormous ethical responsibility on organizational leadership. Nothing can seem more compelling in the wake of a highly public incident or accident than to find a local explanation that can be blamed, suspended, charged, convicted. The problem and the pressure it generates are then simply packed off, loaded onto somebody who can leave, slide out of view, or get locked up—taking the problem along. But it is not ethical, and it is not likely to be productive for the organization and its future safety and justness.

THE INJUSTICE IN JUSTICE

Pursuing justice in court will always produce truths and lies, losers and winners (and more losers). Even if a scapegoat eventually gets exonerated, interests will have been lined up against each other in a way that makes any kind of reconciliation really difficult. By treating error as a crime, we ensure that there *always* will be losers, whatever the outcome of a trial. Because it divides people into groups of adversaries, we guarantee that there will always be injustice in justice, whether the practitioner gets off the hook or not. Common interests dissipate, trust is violated, shared values are trampled or ignored, relationships become or stay messed up.

On May 2, 2001, Chris Sherwood was cleared of any blame at the Old Bailey in London when the judge, Mrs. Justice Rafferty, instructed the jury to find him not guilty. There was no evidence of any intention to kill, she argued, other than that he had fired in self-defense. Justice prevailed, at the same time that injustice prevailed. What about the cover-up during the aftermath? And what about the victim's family or his girlfriend, the woman who had stumbled upon Ashley's still-warm corpse? No recourse for them, no justice, popular opinion found. Many commentators cried foul. The reaction meant that the pressure for the Police Complaints Authority to show its teeth, and for others to charge and convict, would probably remain.

No just culture—no peace for those who do the work every day.

Case Study: When Does a Mistake Stop Being Honest?*

When does a mistake stop being honest? There is no clear a priori line. It isn't as if it would be patently clear to everybody involved that there is a line that has been crossed, and once it has, the mistake is no longer honest. Rather, it depends very much on how the story of a mistake is told, and who tells it. Here I want to recount one such "mistake," a mistake that got turned into a crime, first a "crime" of sorts by the organization itself, and then by a prosecutor (who, in this particular country, happens to work for the regulator). Who gets to draw the line? Who gets to tell the story? These questions are absolutely critical in understanding how we end up with what might be seen as unjust responses to failure.

The incident happened on the morning of November 21, 1989, when a Boeing 747 on an instrument approach in heavy fog came very near to crashing at London Heathrow Airport. The big airliner had been misaligned with the runway so that, when it started pulling up for a second go, it was actually outside the airport's perimeter fence, and only about 75 feet off the ground. It narrowly missed a nearby hotel, setting off car alarms all over the parking lot and fire sprinklers in the hotel. The second approach and landing were uneventful. And most passengers had no idea they had been so close to a possibly harmful outcome. For them, the flight was now over. But for the captain, a veteran with 15,000 flight hours, it was only the beginning of a greater drama. Two and a half years later, a divided jury (10 to 2) would find him guilty of negligently endangering his aircraft and passengers—a criminal offense. He would lose his job, and then some.

I reanalyze an account of the incident as masterfully told by Wilkinson[15] to illustrate the tension and distance between different interpretations of the same event. Was it a mistake culpable enough to warrant prosecution? Or was it normal, to be expected, all in a day's work?

We can never achieve "objective" closure on these questions. You can only make up your own mind about them. Yet the case raises the fundamental issues about a just culture: How can we produce explanations of failure that both satisfy demands for accountability *and* provide maximum opportunities for organizational learning?

A BUG, WEATHER, AND A MISSED APPROACH

Wilkinson describes how the captain's problems began at a Chinese restaurant in Mauritius, an island in the Indian Ocean off Africa. Together with his flight deck crew, a copilot and flight engineer, he dined there during a layover before flying on to Bahrain and then to London. The leg from Bahrain to London would be the last portion of a trip that had begun in Brisbane, Australia.

---

* This case study is taken from Ref. 15.

Several days later, when the flight had gotten to Bahrain, both the copilot and flight engineer were racked with gastroenteritis (stomach flu). The captain, however, was unaffected. A Mauritian doctor had given the flight engineer's wife tranquilizers and painkillers. She also was on the trip and had dined with the crew. The doctor had advised the flight engineer to take some of his wife's pills as well if his symptoms got worse. Flight crews, of course, can't just take advice or prescriptions from any doctor, but this man had been suggested by an airline-approved physician, who was too far away but had recommended the examining doctor to the crew. He would soon be added to the airline's list anyway. He did not, however, seem concerned that the crew had been scheduled to fly in a few days' time again. A colleague pilot commented afterward:

> This was apparently a doctor who didn't even understand the effects of self-medication in a pressurized aircraft on the performance of a complex task, and right there is a microcosm of everything that pressured the crew to get the job done. That doctor's vested interest is in sending flight crews out to fly. Certainly if he ever expects to work for the airline again, he isn't going to ground crews right and left. The company *wants* you to fly. (As part of the court case years later, however, the captain would be accused of violating the company's medical procedures.)

The subsequent flight to London was grim. Unexpected headwinds cut into the 747's fuel reserves, and the copilot had to leave the cockpit for several hours after taking some of the flight engineer's wife's medicines to control his symptoms. It left the captain to fly a stretch of five hours alone, much of it in the dark.

LONDON FOG

Over Frankfurt, the crew heard that the weather at London Heathrow Airport was bad. Thick fog meant that they probably would have to execute a so-called Category III instrument approach. In Category III conditions, a 747 is literally landing blind. While the wheels may just have poked out of the fog in the flare, the cockpit, considerably higher, is still in the clouds. Category III approaches are flown by the autopilot, with the crew monitoring the instruments and autopilot performance. The autopilot captures two radio beams (a localizer for lateral and a glideslope for vertical guidance). These are transmitted by the instrument landing system on the ground, and the autopilot translates them into control commands to make the aircraft stay on track and on a gradual descent, exactly toward and onto the runway. At least, that is the idea.

The captain, like most airline pilots, had never flown a Category III approach down to minimums, despite his extensive instrument experience. The copilot, new with the airline, hadn't either. He had not even had the mandatory training for a Category III approach, and was not approved to fly one. But that was not going to stop anything. Still over Germany, the captain had gotten in touch with the airline and requested permission for the copilot to help out on this one approach into London to get them home. Dispensation was granted. It routinely is. It almost always is. The captain, however, never volunteered that his copilot was not in the best of states (in fact, he may not have been in the cockpit at that very moment). Nobody on the ground inquired either.

Later, the copilot testified that nobody had asked him if he wanted a dispensation. But even if he'd been asked, it would have been difficult to refuse. "I accepted, with the airline's interests at heart, the dispensation to operate to category III autoland conditions," he later wrote to the court. "I personally would not mind if we had diverted. But what would the airline have said to the captain if he had diverted without asking for a dispensation? What would they have said to me if I had not accepted it?"

He really had been in a bind. Wanting to help the airline, wanting to get its passengers home, the copilot had agreed to go on with the flight. But he was sick, really. So if the flight would have had to divert because he was feeling too poorly to do a Category III approach, this once, what was he doing on board anyway? And where had those medicines come from?

"This," Wilkinson, observed, "is the heart of the professional pilot's conflict. Into one ear the airlines lecture, 'Never break regulations. Never take a chance. Never ignore written procedures. Never compromise safety.' Yet in the other they whisper, 'Don't cost us time. Don't waste our money. Get your passengers to their destination—don't find reasons why you can't.'"

THE APPROACH

Nearing London, the 747 was given a routine holding northeast of the airport. After some time of flying racetracks in the holding pattern, the flight engineer suggested, "Come on, we've got two minutes of holding fuel left, let's buzz off to Manchester." The crew discussed the options—both Manchester and Gatwick (south of London) were diversion airports, though Manchester had better weather. But the captain "was a very determined man," as the flight engineer recalled. Just as he was deciding to head off to Manchester, Heathrow called and cleared the 747 for approach.

But a complication had arisen: instead of landing to the east (runway 09), as had been planned, they would now have to turn in shorter and land toward the west (runway 27), because the wind had changed. The approach became a hurried affair. The crew had to reshuffle charts, think and talk through the procedures, revise their mental pictures. A 10-knot tailwind at altitude meant that the 747 was motoring down the approach path toward the runway at an even greater groundspeed. Tightening their slack further still, the approach controller turned the 747 onto the localizer 10 miles from the runway, rather than the normal 12 miles or more. Halfway down, the tower radioed that some approach lights were apparently not working, requiring the flight engineer to take a quick look through his checklist to see how this, if at all, affected their planned procedure. The tower controller also withheld clearance for the 747 to land until the last moment, as a preceding 747 was feeling its way through the fog, trying to find its turnoff from the runway.

But the autopilots really were about to become the final straw: they never seemed to settle onto the localizer, instead trundling back and forth through the beam, left to right. The two autopilots on this old, "classic" 747 may never have been able to capture the localizer: when the aircraft turned in to start its approach, the autopilots disconnected for some time and the airplane was flown manually. The autopilots, built by Sperry, were based on an earlier design. They were never really meant for this aircraft, but sort of "bolted on," and had to be nursed carefully. On this flight the crew made a later attempt to reengage the autopilots, though radar pictures showed that the 747 never settled on a stable approach path.

The flight engineer was getting worried about the captain, who had basically been flying solo through the night, and still was alone at the controls. The copilot was of little help. "I was not qualified to make this approach and could not make any suggestions as to what was wrong," he would later tell safety investigators. He stayed out of the way.

The captain was now technically illegal: trying to fly a Category III approach with autopilots that refused to settle down and function perfectly was not allowed. The right decision, to everybody in hindsight, would have been to go around, to fly what's called a missed approach. And then to try again or go to the alternative. "I'd have thrown away the approach, gone to my alternate or tried again. No question about it," one pilot questioned by Wilkinson said.

But other pilots, some with the same airline, believed the opposite. "Look, he was concerned about fuel. He had a first officer who was no help. He knew a diversion to Manchester would cost the airline a minimum of $30,000. He realized he'd be sitting in the chief pilot's office trying to explain how he got himself into a position that required a missed approach in the first place. He figured the autopilots would settle down. And I'll bet he was convinced he'd break out at Category I limits (a higher cloud ceiling and better visibility than Category III) and could take over and hand-fly it the rest of the way. I can understand why he carried on."

It might have worked, Wilkinson observed. And if it had, nobody would ever have heard of this case.

But it did not work. Ever concerned with passenger comfort, the captain waited with making a go-around. And then he made a gentle one. The 747 sank another 50 feet. The flight engineer glimpsed approach lights out the left window as they started pulling up, away.

As one 747 instructor said, "This is a pilot who was critically low on fuel, which probably was one reason why he waited a second before going around. At decision height on a Category II approach, you look to see the slightest glow of approach lights, you wait 'one-potato,' see if anything comes into sight. Perhaps a thousand times before, he'd watched that same autopilot do strange things on the same approach to the same airport, and he'd break out at two hundred or five hundred feet and make a play for the runway. And on the crew bus everybody says, 'Boy, that autopilot sucked again today.'"

On climb-out after the first try, the copilot noticed how the captain's hands were shaking. He suggested that he fly the second approach instead, but the captain waved him away. The second approach was uneventful, and was followed by a landing that elicited applause in the passenger cabin.

NO DISCLOSURE, BUT A TRIAL

Back in the crew room after they had shut down the airplane, the captain found a note in his company letterbox. It requested that the crew see the chief pilot. The captain told the copilot and flight engineer to go home, and he would say that they had already left when he found the note.

But he did not go to the chief pilot either. Nor did he talk to an airline safety investigator about what had happened. Instead, he drove straight home and went to bed. That evening, a call came from the airline. The crew had been suspended.

An internal investigation was launched by the airline, who later issued a report chiding the copilot and flight engineer. The airline also demoted the captain to first officer. The aviation authority downgraded his license accordingly, and he was relegated to riding out the rest of his career in the right seat, no longer in command.

This was too much. Half a year after the incident, the captain resigned from the airline and began to appeal the authority's reduction of his license. Some did not see any problem. Recently, the pilot had been receiving grades of "average" on his half-yearly proficiency checks in the simulator, and instructors had taken note of his inability to perform well under pressure.

But why did the regulator take him to court? This "remains the subject of speculation," Wilkinson writes. "There is considerable feeling that the airline was not sorry to see it happen, that the captain was a loose cannon who could have made things awkward for an airline that places great value on its public image. Some feel that the captain could have revealed some controversial company procedures. If the captain were branded a criminal, it would effectively negate whatever damage he might do... Others suspected empire building within the regulator's legal branch: this looked like a juicy case for an aspiring prosecutor to take public and demonstrate that even the flag carrier's jumbo jet captains dare not take on the aviation authority casually."

Six weeks after the incident, the airline had announced that it was no longer granting bad-weather dispensations. But the fleet manager who had authorized the approach with the copilot's dispensation was not in the dock. Nor was the controller who turned the big 747 onto a tight approach, separated by what seemed like only five miles rather than the legal minim of six from the preceding 747. With traffic from all over the world converging onto London at eight in the morning, those rules were obviously allowed to be flexible.

It was the pilot who was in the dock. Seated next to a policeman. Why had he not filed a Mandatory Occurrence Report right after the flight? Because it did not constitute an occurrence, the pilot argued. After all, he had gone around, or at least initiated a go-around, and landed uneventfully the second time. Why had he gone around so slowly? Because the supposedly canonical technique was not described anywhere, he argued. At some point in the trial, the pilot produced a transcript of every oral callout, checklist response, and radio transmission that company and government regulations required the crew to accomplish during the approach. It showed that the entire routine took seven minutes. The approach had lasted only four, making it technically impossible to make an approach and follow all applicable rules at the same time.

Few cared. Jurors sometimes even napped. If the trial did not revolve around arcane legal points, it did so around finely grained technical ones. The pilot was never called to testify on his own behalf.

The defense elaborated the fact that the old 747 was dispatched on its next leg out of London without a check of the autopilot, to see if it was somehow faulty. To this day, four crucial pages of the maintenance log, which might have told something about the autopilot, are missing (in a parallel to the prescription missing from the medication log in Mara's case; see Chapter 3).

"The regulator itself was at fault," a legal expert and airline pilot commented, "for permitting a situation to exist in which the airline's flight operations manual contained a provision that the captain would be expected to use, by which it could authorize him to make the approach without a qualified copilot. The approach was actually illegal at the fault of the airline, yet they were not charged. Had that provision not existed, the captain would have diverted to Frankfurt with cozy fuel reserves, to await better weather at London."

A split jury found the pilot guilty. The judge fined him only £1500 and rejected the regulator's demand that he pay £45,000 more to cover court costs. The pilot appealed the decision, but that was summarily rejected.

When he was young, the pilot lived near an air force base where he would watch airplanes take off and land at the end of the war. That inspired him to become a pilot. "On December 1, 1992, three years and nine days after the incident, the pilot left home without a word to his wife. He drove some nine hours to a beach near the air force base. There he ran a hose from his car's exhaust pipe through a nearly closed window. In a matter of minutes he was dead. He left no letter or any explanation."

It would be too easy to ask whether the prosecution and conviction of the captain was right. Or just. Because it is too difficult to answer. Was this a crime?

Multiple descriptions of the events are plausible. The disappearance of documents without a trace in these cases can always give people the chills. Was it a conspiracy after all, a "cover-up," as some of Wilkinson's interviewees suggested? It could have been: turning one pilot into a highly visible scapegoat to silence him and others. This would save the reputation of both the airline and the regulator, who also happens to employ the aviation prosecutor in this country. But conspiracies take tight coordination and demand iron discipline from those involved in them.

Also, as a captain, this pilot had lately been "average," not stellar. He was stubborn and determined. He was ultimately responsible for getting himself and his crew into this jam. And then he apparently refused to cooperate and did not want to disclose or discuss the incident (it wasn't an occurrence to him, after all) until forced to do so in the adversarial setting of a trial.

Who is right? Whose version of event is true? The tension between multiple possible interpretations remains until the end of Wilkinson's story. But important points about building a just culture do stand out.

- **A single account cannot do justice to the complexity of events.** Like the physicist Niels Bohr when he tried to convince his colleagues at the time—Einstein and Heisenberg—we need multiple layers of description, partially overlapping and always somehow contradictory, to have any hope of approximating a rendition of reality.
- **A just culture accepts nobody's account as "true" or "right"** and others as wrong. This only leads to moral grandstanding, imperialism, and to losing situations, like this pilot's. Instead, it accepts the value of multiple perspectives, and uses them to encourage both accountability and learning.

- **A just culture is not about absolutes, but about compromise**. Achieving justice is not about black and white. Instead, it presumes compromise. Justice in a just culture cannot be enforced; it must be bargained. Such bargaining for justice is a process of discovery, a discovery that the best bargain may be an outcome in which every party benefits, for example, an explanation of events that satisfies calls for accountability *and* helps an organization learn and improve.
- **A just culture pays attention to the "view from below"** among these multiple accounts, as that view (in this case from the person in the dock) may have little or no power to assert itself and is the easiest to quash. Silencing it can be organizationally or politically convenient. You may even see it as imperative. You may see putting others in an inferior position as a necessary, if sometimes annoying step in achieving other goals. But this makes it even more morally essential to give the view from below a voice.
- **A just culture is not about achieving power goals**, by using other people to deflect attention away from one's own flaws. This denies such people their personhood; it makes them a mere instrument in the pursuit of protection of power, of existing structures or arrangements. Most people will see this as unethical, and it violates the basic principles of Aristotelian justice that many of our societies still live by.[16]
- **Disclosure matters**. Not wanting to disclose can make a normal mistake look dishonest, with the result that it will be treated as such. Multiple examples in this book illustrate this. Disclosing is the practitioner's responsibility, or even duty.
- **Protecting those who disclose matters just as much**. The demand to disclose in the pilot's case above (a note in the letterbox) may not have given him confidence that honest disclosure would be treated fairly. Conditions at his airline may have been unfavorable for honest disclosure. Creating a climate in which disclosure is possible and acceptable is the organization's responsibility. And more protections are often necessary.
- **Proportionality and decency are crucial to a just culture**. People will see responses to a mistake as unfair and indecent when they are clearly disproportionate. "What was the guy found guilty of?" a pilot friend had asked Wilkinson in amazement. "Endangering his passengers," Wilkinson replied. "I do that every day I fly," the friend said with a laugh. "That's aviation."[15] The eventual punishment given to this pilot (a symbolic fine) may have indicated that the trial was seen as a disproportionate response to an event that perhaps should not have ended up in court. Proportionality means heeding Martin Buber's dictum: "What is necessary is allowed, but what is not necessary is forbidden."

By the time a case reaches trial, much of the preceding has either been wasted or rendered impossible. A trial cannot do justice to the complexity of events, as it necessarily has to pick one account as the truest or most trustworthy one.

A seeming lack of honest disclosure is often a trigger for a trial. This could have been the case here. You can also see it in the literature on medical lawsuits. Patients or their families do not typically take a doctor to court *until* they feel that there is no longer any other way to get an account of what went wrong.[17] Stonewalling often leads to a trial. But a climate that engenders anxiety and uncertainty about how disclosure will be treated often leads to stonewalling. The more we take cases to trial, the more we could be creating a climate in which freely telling each other accounts is becoming more and more difficult.

1 Retributive and Restorative Just Cultures

There are basically two ways to approach a just culture: inside your organization as well as outside of it. One is based on retribution, the other on restoration. When faced with actions that led to a (potentially) bad outcome

1. **Retribution** imposes a deserved and proportional punishment.
2. **Restoration** repairs the trust and relationships that were damaged.

Retributive and restorative processes ask very different questions in the wake of an incident. Consider this when you are trying to build a just culture of trust and accountability in your organization.

- Retribution
  - Which rule has been broken?
  - Who did it?
  - How bad was the infraction, and so what does the person deserve?
  - Which manager, department, or authority gets to decide?
- Restoration
  - Who has been (potentially) hurt?
  - What are their needs?
  - Whose obligation is it to meet those needs?
  - What role does the community play in learning from the event?

Let's first look at retributive and restorative approaches to justice separately, to see how each tries to generate accountability, trust, and learning. Then we'll compare and contrast the two. You will find that the two approaches don't have to be mutually exclusive. In your own organization, you might well be able to create a mix of responses that imports the best of both worlds.

RETRIBUTIVE JUST CULTURE

Retributive justice is as old as humanity itself. It is easily recognizable in varieties of ancient law (e.g., the law of "measure for a measure"). Retribution considers punishment, if proportionate, to be a just response to a sanctionable action and to be an appropriate deterrent. Retributive justice wants the offender to pay something back, to forfeit something in return for what he or she did. Retribution is not vengeance. Retribution is not supposed to be directed at the person, but at the actions (the wrongs). It is supposed to come with inherent limits (i.e., there is only so much punishment that you can exact), it involves no pleasure at the suffering or pain of others, and it is constrained and governed by procedural standards.

Shades of Retribution

A popularized way to think about a just culture is based on shades of retribution. This type of just culture, introduced in hospitals, airlines, oil companies, and other organizations over the past years, understands that it is foolish to expect fallible people to be perfect. These organizations find it important to learn from mistakes, and their approach to just culture wants to hold people accountable not necessarily for the outcomes they create, but for the choices they (supposedly) make while doing their work. When things go wrong, the consequences should depend on the choices that people made in the lead-up. Such a just culture often distinguishes among a minimum of three types of actions:

- An **honest mistake** is an inadvertent lapse, slip, or mistake. It was unintended and can happen to anyone in those circumstances.
- **At-risk behavior** is a choice that increases a risk that is not recognized, or mistakenly believed to be justified.
- **Negligence or recklessness** is a choice to consciously disregard or take a substantial and unjustified risk.

Those who have adopted such a "just culture" believe that different choices deserve different consequences—different shades of retribution. Honest mistakes require compassion and an investigation into the conditions that triggered them. At-risk behavior calls for coaching and warnings. Recklessness must lead to disciplinary action, including suspension, dismissal, or referral to other authorities. The idea is that employees who feel that they will be treated fairly—that their honest mistakes will not be punished—are more inclined to report mishaps and failures. But will they?

"You have nothing to fear if you have done nothing wrong."

What if a prosecutor would say this to respond to concerns from the community of practitioners? It might be said to assuage concerns from the sector that human errors—normal, honest mistakes—are being converted into criminal behavior by the prosecutor's office. This is not just hypothetical, of course. Practitioners have been fined or charged for rule infractions that were part and parcel of getting the job done. They and their colleagues may feel some anxiety as a result. Can they supply incident data in good faith, or are the data going to be used against them? Are there enough protections against the prying of a prosecutorial office? Don't worry, a prosecutor might say in reply. Trust me. There is nothing to fear if you have done nothing wrong. I can judge right from wrong. I know a willful violation, or negligence, or a destructive act when I see it.

But does he? Does anybody?

Retributive just cultures draw a line somewhere between acceptable and unacceptable behavior. A willful violation is not acceptable. An honest mistake is.

Risk-taking behavior probably isn't. Negligence and recklessness are certainly not. And if what you have done is acceptable—if you have done nothing wrong—you have nothing to fear. A just culture definition in use in air traffic control, for example, promises and warns simultaneously that "Front-line operators or others are not punished for actions, omissions or decisions taken by them that are commensurate with their experience and training, but where gross negligence, willful violations, and destructive acts are not tolerated."[18]

The idea of a line makes intuitive sense. If just cultures are to protect people against being sanctioned for honest mistakes (when they've done nothing wrong), then some space must be reserved for mistakes that are not "honest" (in case they *have* done something wrong). Consequently, all proposals for a just culture emphasize the establishment of, and consensus around, some kind of line between legitimate and illegitimate behavior: "In a just culture, staff can differentiate between acceptable and unacceptable acts."[19] There is not a single proposal for just cultures—indeed, not a single appeal to the need to learn from failure in aviation—that does not build in some kind of escape clause into the realm of essentially negligent, unwanted, illegitimate behavior. An environment of impunity, the argument continues, would neither move people to act prudently nor compel them to report errors or deviations. After all, if there is no line, then "anything goes." So why report anything? This is not good for people's morale, for the credibility of management, or for learning from mistakes and near misses.

So calls for some kind of border that separates tolerable from culpable behavior make intuitive sense. And ideas on just culture often center on its embrace and clarity: "A 'no-blame' culture is neither feasible nor desirable. Most people desire some level of accountability when a mishap occurs. In a Just Culture environment the culpability line is more clearly drawn."[13] Another argument for the line is that the public must be protected against intentional misbehavior or criminal acts, and that the application of justice is a prime vehicle for such protection.

A recent directive from the European Union (2003/42/EC) governs occurrence reporting in civil aviation. This directive has a qualification: a state must not institute legal proceedings against those who send in incident reports, apart from cases of gross negligence. But who decides what counts as "gross negligence"? The same state, of course. Via its prosecutors, investigating magistrates, and judges.

The directive, as does much guidance on just culture today, seems to assume that cases of "gross negligence" jump out by their very nature, that "willful violations" represent an obvious category, distinct from violations that are somehow not "willful." It assumes that a manager, a prosecutor, a judge, or any other authority can recognize—objectively, unarguably—willful violations, negligence, or destructive acts.

If we want to draw a line, we have to be clear about what falls on either side of it. Otherwise there is no point in a line—then the distinction between acceptable and unacceptable behavior would be one big blur. Willful violations, say many people, clearly fall on the "unacceptable" side of the line. Negligence does too. But what is negligence then? To begin with, look at this definition:

> Negligence is conduct that falls below the standard required as normal in the community. It applies to a person who fails to use the reasonable level of skill expected of a person engaged in that particular activity, whether by omitting to do something that a prudent and reasonable person would do in the circumstances or by doing something that no prudent or reasonable person would have done in the circumstances. To raise a question of negligence, there needs to be a duty of care on the person, and harm must be caused by the negligent action. In other words, where there is a duty to exercise care, reasonable care must be taken to avoid acts or omissions which can reasonably be foreseen to be likely to cause harm to persons or property. If, as a result of a failure to act in this reasonably skillful way, harm/injury/damage is caused to a person or property, the person whose action caused the harm is negligent.[13]

A few concerns come to mind. First, the definition is long. Second, it is actually not really a definition. It does not capture the essential or finite properties of "negligence." It does not allow you to grab "negligent behavior" and put it on the unacceptable side of the line. Instead, it presents an array of questions and judgments that you need to make. Rather than this solving the problem of what is "negligence" for you, you now have to solve a larger number of perhaps equally intractable problems instead:

- What is the "normal standard"?
- How far is "below"?
- What is "reasonably skillful"?
- What is "reasonable care"?
- What is "prudent"?
- Was harm indeed "caused by the negligent action"?

So instead of clarifying which operational behavior is "negligent," this shows just how complex the issue is. And how much of a judgment call it is. In fact, there is an amazing array of judgment calls to be made. Just see if you, for your own work, can (objectively, unarguably) define concepts such as "normal in the community," "a reasonable level of skill," "a prudent person," or that you could achieve "a foresight that harm may likely result." What, really, is normal (objectively, unarguably)? Or prudent, or reasonable (objectively, unarguably)? Don't we all want to improve safety precisely because the activity we are engaged in can result in harm? And, of course, research showed us a long time ago that once we know the outcome, we overestimate the amount of foresight we, or someone else, could and should have had.[20]

Just responses to bad events are not a matter of matching the inherent properties of undesirable behavior with appropriate pigeonholing and a fitting punitive level. It involves the hard work of deciding what story to tell and whether to see something as reckless, as at-risk, or as erroneous. Merely supplying the categories leaves this issue unresolved. It boils down to fairly empty guidance on how to create a just culture. It also creates an ethical issue: assigning an act to a category will forever be a judgment.

It is not that making such judgments is impossible. In fact, we do this often. It is, however, important to remember that these are indeed judgments. They are not objective and not indisputable. To think that there comes a clear, uncontested point at which everybody says, "Yes, now the line has been crossed, this is negligence," is an illusion. What is "normal" versus "negligence" in a community, or "a reasonable level of skill," versus "recklessness" is infinitely negotiable. You can never really close the debate on this. As a result, there really is no line. There only are people who draw it.

What matters is not whether some acts are so essentially negligent as to warrant serious consequences. What matters is which authorities we in society (or you in your organization) rely on to decide whether acts should be seen as negligent or not. Who draws your line?

Assigning acts to categories becomes a matter of power. Who has the power to tell the story, to say that behavior is one thing and not the other? And who has the power to decide on the response? It is a power that can finesse and fudge a whole range of organizational, emotional, and personal issues. A conclusion of wrongdoing could, for instance, be underwritten by a hospital's risk manager's greatest fears (of liability, loss of reputation or political clout) or by how a manager is held accountable in turn for evidence of trouble in the managed unit.

Suppose nurses take to scanning the barcode label that one of their colleagues pasted on the wall behind the patient because it actually reads well, and is always easy to find as opposed to others. This may have become all but normal practice—everybody does it because everybody always has the next patient, and the next, and the medication barcode scanners are of such poor quality that they can't read anything except flat and high-contrast labels (indeed, labels pasted on a wall).

Managers may want to call such behavior "at-risk" and mete out supposedly appropriate countermeasures. But nurses may no longer see their behavior as at-risk, if they ever did. In fact, it may be a sure way to get a good scan, and not doing that could create more risks. And, of course, barcode scanning is not their main job—taking care of patients is. This means that nurses may see any punitive responses to scanning a label stuck on a wall as pretty unjust. After all, such responses may show that the manager is not aware of the unrelenting pressures and ebbing and flowing demands of nursing work, or of the shortcomings of putatively labor-saving and safety-enhancing technology. Justice is a matter of perception.

But managers are under different pressures. Managers appropriate the power to call something "at-risk" not because of their putative insight into the risks and realities of nursing work, but because they can and because they have to relative to the pressures and expectations of their own positions in the hospital. From a manager's point of view, operational behaviors that bypass instructions or protocol, for example, could end up eroding productivity and reputation and eventually impair the financial performance of his or her part of the organization. Or, for that matter, having to make structural or equipment changes (e.g., procuring new or better barcode scanners) involves sacrifices that are much larger than reminding people to be more careful and follow the rules.

What if we involve domain expertise in drawing the line? Or get the experts to do it altogether? Does that help? There is no research that suggests that domain experts automatically prevent the biases of hindsight slipping into their judgments of past performance. Hindsight is too pervasive a bias. And domain experts may have additional biases that work against their ability to judge the quality of another expert's performance fairly. There is, for example, the issue of psychological defense: if experts were to affirm that the potential for failure is baked into their activity and not unique to the practitioner who happened to inherit that potential, then this makes them vulnerable too. Sometimes it can be more comforting to think that the errors made by a fellow practitioner would not happen to you, that they really are unique to that other person.

Difficulties and Fairness in Retribution

Could retributive just culture approaches undermine honesty and reporting and lead to arbitrary judgments after all? To date, no research evidence has been offered that organizations that have implemented a retributive just culture program are better at learning (as indicated by a greater willingness to report safety events or incidents). That said, there is only limited evidence that retributive just culture approaches inhibit openness and learning.

1. Systematic evidence comes from a 2006 survey study (n = 1984) into perceptions of just culture across disciplines in healthcare. In exploring the existence of retributive responses, it asked questions such as, "Are employees held accountable for their actions?" "Is there blame or favoritism?" "Does the organization recognize honest mistakes?" Accountability was perceived significantly differently. Physicians tended to have the most positive view of their culture as "just," followed by management—over groups with less power in the medical competence hierarchy (e.g., nursing and non-clinical staff), who considered the same culture significantly less "just."[21]
2. Anecdotal evidence suggests that a program that "blames workers themselves for job injuries and illnesses, drives reporting underground. If injuries aren't reported, the hazards … go unidentified and unaddressed."[22]
3. Evidence by analogy. In a study of workplace complaint systems, formal options that led to retribution and top-down responses affected people's willingness to disclose. If "the organizational culture is too hierarchical and oriented toward punishment, [this] may inhibit willingness to act or come forward."[23]

Retributive just culture approaches are organized around finding out *who* did something wrong and how to deal with *him or her*, rather than asking *what* was responsible for things going wrong and how to fix *that*. If the response focuses on the individual, the who, then underlying conditions that gave rise to the problem in the first place can be missed and left unaddressed. In one case, a nurse was blamed for the death of a 16-year-old patient, dismissed from her job, and then criminally prosecuted.[24] A later report, however, found how systemic issues at the hospital set the stage for the nurse's fatal drug error.[25] Asking *who* was responsible downplayed the role played by other factors—local pressures and constraints; scheduling issues; fatigue; personnel shortages; systematic gaps in support; barcode scanner problems; issues of training, hierarchy, and information flow; and officially condoned workarounds. By blaming the nurse, nothing of value was learned from the patient's death.

Let's look at two important aspects of retributive justice here, and see how these play out in a retributive just culture.

1. **Substantive justice** prescribes how regulations and rules that people are supposed to follow in their jobs must themselves be fair and legitimate.
2. **Procedural justice** sets down legitimate processes for determining rule breaches, offers protections for the accused, and governs who should make such determinations.

Substantive Justice

Substantive justice relates to the morality and legitimacy of a rule's content. If rules themselves are unfair, illegitimate, or obviously contradicted by other workplace expectations, then there is little point in trying to have a "just culture" response to their violation. It won't ever be just.

In the case of the 16-year-old patient and nurse mentioned previously, a workaround "rule" had been approved by the hospital in which anesthetists could give drug orders by phone from the floor where they worked and did not have to see patients in the obstetric ward at that time. Nurses could do all the actual work by the bedside. When things went fatally wrong this time, it wasn't the workaround or unfairness of the rule that was put on trial: it was the nurse. You might say, of course, that workers like that nurse consent to organizational (and other) rules when they take up their occupations. In many cases, such consent is indeed an important consideration in assessing the rule's legitimacy. But consent is not definitive. Workers can be placed in positions where they have little choice but to submit to preexisting rules or workarounds.

Rules may not be written with everybody's legitimate interests in mind, or in ignorance of the goal conflicts and resource constraints that drive real work. Here is an example from Ref. 26.

Breaking the Rules to Get More Recruits: Some Say Cheating Needed to Fill Ranks

It was late September when the 21-year-old man, fresh from a psychiatric ward, showed up at a US Army recruiting station. The two recruiters there quickly signed him up. Another recruiter said the incident hardly surprised him. He has been bending or breaking enlistment rules for months, he said, hiding police records and medical histories of potential recruits. His commanders have encouraged such deception, he said, because they know there is no other way to meet the Army's recruitment quotas.

"The problem is that no one wants to join," the recruiter said. "We have to play fast and loose with the rules just to get by." Others spoke of concealing mental health histories and police records. They described falsified documents, wallet-size cheat sheets slipped to applicants before the military's aptitude test, and commanding officers who look the other way. And they voiced doubts about the quality of troops destined for combat duty.

Recruiting has always been a difficult job, but the temptation to cut corners is particularly strong today, as deployments in Iraq and Afghanistan have created a desperate need for new soldiers, and as the Army has fallen short of its recruitment goals in recent months. Says one expert: "The more pressure you put on recruiters, the more likely you'll be to find people seeking ways to beat the system."

A retributive just culture presumes that existing rules enjoy a priori legitimacy. But what if they don't, or when they are really trumped by other organizational pressures and expectations as in the example above? Listening to the voice from below—from those who have to get the job done and use the rules—is the best way to achieve substantive justice.

- Involvement of those who will have to do the work enhances the legitimacy of the rules that apply to that work.
- Taking part in the process of developing the rules increases the sense of ownership the workers feel toward the rules. The rules derive from their own insights, arguments, and experiences.
- Developing the rules in connection with the workers ensures the rules connect with reality. The standards are not designed for an ideal environment, imagined without time pressures, complicating factors, and conflicting information. Instead, the written rules (and practices taught by educators) align with and support normal practice in the field.

Procedural Justice

Procedural justice sets down legitimate processes for determining rule breaches, offers protections for the accused, and governs who should make such determinations. Let's look at

- The necessity for independent judges
- The right to fair hearing and appeal
- A differentiation between guilt-phase and penalty-phase deliberations

Independent judges have no personal stake or conflict of interest in the affair at hand. This aspect of procedural justice bars a person from deciding a case in which she or he has something to win or lose. Think about what this means for a line manager who applies a just culture process to one of his or her reports. Line managers can be fairly suspected (if not shown) to have reputational, career-related, economic, or other stakes in adjudicating an error or violation that happened on their watch. This gives them an interest in the outcome, which introduces the kind of actual or suspected bias that procedural justice and due process rights aim to prevent. At the same time, knowledge of the messy details or subtleties of what it takes to get a job done under goal conflicts and resource constraints is certainly as important. Research has shown that the legitimacy of being called to account is linked to how much the judging person knows about the process, profession, or practice in question.[27] Those who know the "messy details" of real (rather than imagined) work tend to enjoy greater credibility. The problem of course is that these are not likely to be independent. Finding a judge who is both independent and intimately knowledgeable about how work is actually done may require an organization to look outside or across different sites or parts of itself (such as employing another line manager from a separate site as an independent assessor).

Due process rights also include the right to a fair hearing. This normally involves

- Prior notice of the case made against you and getting to see the evidence that will be used against you.
- Knowing what is at stake (not just that there is "a case").
- A fair opportunity to answer the case and evidence against you.
- The right to bring an advocate who might support you or argue on your behalf (this could be a colleague or union representative).
- An opportunity to present your own case or angle.
- Openness to scrutiny of the case and its proceedings by other parties. If things are done behind closed doors, then this opportunity may not exist. If colleagues or other stakeholders are present, then it does.
- The right of appeal. This offers a process for requesting a formal change to an official decision. Appeal may be called for, among other reasons, because of a suspicion of abuse of power, because someone acted in excess of jurisdiction, because evidence was used that shouldn't have been included, or because evidence was ignored that should have been considered.

Nothing in a typical just culture program itself offers such assurances, even though it is possible (and practically and morally desirable) to develop and offer them. If you have a retributive just culture process, then think what it would take in your own organization to give your employees assurance of procedural justice. When you see a list of procedural guarantees like the one above, you might realize that having a just culture is no longer so simple. It is certainly not as simple as buying an algorithm with three categories (honest mistake, risk-taking behavior, negligence) off the shelf.

If you are ready for even more nuances and complexities, then consider this. Even with procedural guarantees in place, a "just culture" program organized around categories of culpability does not distinguish between guilt-phase versus penalty-phase culpability. This divides the question about culpability in two:

1. Did the person knowingly commit the act (guilt phase)?
2. What penalty should be assigned once guilt is established (penalty phase)?

A retributive just culture program presumes guilt, or at least (some) responsibility for the outcome; otherwise the algorithm wouldn't be applied. This might remove the presumption of innocence until proven otherwise, and thus create a short circuit to penalty deliberations. Research has shown that penalty-phase deliberations often focus on the transgression, the transgressor, and the outcome, rather than on mitigating factors.[28] A just culture process that makes no differentiation between guilt-phase and penalty-phase deliberations might automatically overlook mitigating factors (the-*what*-is-responsible question) in favor of asking who is responsible and holding that individual accountable.

These are all checks and balances that have made it into retributive justice throughout the ages. This has occurred under names like "natural justice," "due process," or "the duty to act fairly." Lots of people have fought and died for these assurances. Yet many organizations today that have adopted a retributive just culture are proceeding without any of that in place. Without such checks and balances, your culture will not be seen as just. It will only be seen as an exercise of power, as a serious game that decides who wins and who loses.

Summarizing and Managing the Difficulties with Retributive Justice

Given the negotiability of the line between acceptable and unacceptable behavior, it is not surprising that a just culture based on retribution typically creates the following problems:

- There is a lack of clarity, agreement, or perceived fairness about who draws the line between the shades of culpability. In many cases, the judge is not independent: he or she actually has a stake in the outcome.
- Does the judge or "jury" know the nuances and messy details of the practitioner's work? If not, how can they really know what constitutes risk or risk-taking in that world?
- There is often no possibility for appealing a decision that is made. This type of retributive just culture is known by some employees "as a good way to get yourself fired" and by managers "as a good way to get rid of someone."
- There is actually no convincing evidence that organizations with a retributive just culture have higher reporting rates or that they learn more of value after an incident.
- The more powerful people in an organizational hierarchy typically consider their organization's culture to be more "just."[21]
- Retributive justice is not always known to promote honesty, openness, learning, and prevention.

So if you believe you want to institute a just culture based on retribution, be sure to at least assess these three questions for your own organization:

1. Is the "judge," the one who draws the line on the practitioner's behavior, **independent**? A "judge" who has a stake in the outcome is not independent. For instance, a nurse manager who assesses the performance of one of his or her nurses in case of a medication adverse event is not independent. In fact, any manager is not independent. They always have a stake in the outcome of the judgment, as they and their decisions may be implicated in an incident too—unless blame can be put on the worker.

2. Does the "judge" or "jury" know enough about the **messy details** of practice to know about the many unwritten rules, standards, and expectations about how work actually gets done? Are you allowing people's choices and actions to be judged by their "peers"?
3. Is there an **opportunity for appeal**? If you really acknowledge that everybody is inescapably fallible (as is one of the premises of a retributive just culture), then that goes for judges and juries too. Justice should therefore offer people a chance to be heard again by an unbiased party. Does that happen in your organization, and if not, how could you provide that assurance?

Here are some additional steps and considerations for your organization if you do wish to pursue a path of retributive just culture.

- Design and advertise your just culture process clearly. Tell your people, for example, where and with whom it starts, what the following steps are, where and when a judgment is made on the employee's behavior, and what the opportunities for appeal are.

Recognize that an adverse event review is not a performance review. Consistency of your review processes across professional groups and departments is difficult, but also important for achieving fairness and justice. Training your investigators to conduct the kind of learning review that asks *what* is responsible for an incident (rather than *who* is responsible) is an important part of this.[29]

- Decide who is involved in the just culture process in your organization. If the employee's manager is in charge of the process, then the potential of career or reputational jeopardy may hinder an employee's honest disclosure.[20] A set-up in which impartial staff take in the story and then funnel it to the manager for appropriate action can generate more opportunities for learning and less fear of retribution.[17,30] Yet how much domain expertise do you need to involve in the process? Understanding the messy details of practice (including the many gaps between rules or guidelines and actual daily work) is crucial for both credibility and a sense of justice.
- Decide who is involved in deciding who is involved. There is something recursive about this, of course. But if decisions are made top-down about the previous points, then any just culture process will lack the buy-in, ownership, and constituency for employees to feel that it is something of their own creation—something to stand for and participate in to the benefit of the organization.

I recall how one safety-critical industry was under intense media scrutiny in a country where I once lived. The newly elected government had pledged to the public that it would let the industry continue to function if it were safe. Then reports started to leak out about operators drinking on the job, about an internal erosion in safety culture, about a lack of trust between management and employees. The regulator was under exceptional pressure to do something. To show that it, and the government, could be trusted.

So the regulator sent parts of the cases it had discovered to the prosecutor. The media loved it: now something was happening! Maybe crimes had been committed by people to whom the public had entrusted the running of this safety-critical technology! Now somebody was finally going to be held accountable.

The regulator saw how some of the media spotlight on it got dimmed. It could breathe a little easier now. But it was a bittersweet lull. The relationship with the industry was dramatically disturbed. Regulators have to rely on open disclosure by people in the industry they regulate; otherwise they have no accurate or truthful information to go and regulate on. Such disclosure was now going to be very unlikely. It would be, for years to come.

In addition, safety improvements, at least for the media (and thereby public opinion, and, by extension, the government's stance on the issue) could now be largely collapsed into the pursuit of a few bad apples in the industry's management. Now that these people would be held accountable, any other safety improvements could simply be assumed to be less important, or to follow automatically. Of course they would not. Publicly or legally reminding people of their responsibilities may have some effect in getting them or others to behave differently (though never for a long time). And the negative consequences of such accountability easily outweigh these effects.

RESTORATIVE JUST CULTURE

Restorative justice as we know it today began in response to relatively minor crimes—often property crimes, such as burglary in the 1970s. But its roots can be traced much further back: hundreds and even thousands of years. The ancient Sumerians, Babylonians, Hebrews, Romans, and Gauls—to name a few—all applied forms of restorative justice. They organized restitution for property crimes, relying on input from victims and offenders. First nations in North America, New Zealand, and other parts of the world have also long had values and practices consistent with restorative justice. Encouraged by the results and humane approach, stakeholder groups have advocated restorative justice particularly for juvenile offenses in a number of countries. And it has been spreading to schools, workplaces, and religious organizations. The restorative approach to justice has even been applied successfully on a massive scale, through the Truth and Reconciliation Commission in post–Apartheid South Africa. Restorative justice, broadly, is defined as

> ...a process where all stakeholders affected by an injustice have an opportunity to discuss how they have been affected by the injustice and to decide what should be done to repair the harm.[31]

Restorative justice does not have to be very difficult, actually. And it can even come intuitively, without policies or rules to guide it. Not long ago, I was speaking with hospital administrators in Asia who run a large, complex campus hospital. A delivery man, coming in with supplies on a cart, had knocked over a pram with a baby in it. The mother was of course quite upset, even though the baby was not hurt in the incident. The mother, in conversation with administrators, gave the hospital seven days to come up with an intervention that would prevent it from happening again. She took no other action, neither toward the hospital nor toward the delivery company or its employee.

Once they starting looking into the problem during those days, hospital administrators found that they had no guidance for their deliveries whatsoever: delivery contractors did what seemed right, and there was a wide variety of practices and quality. They developed standards in consultation with their contractors, and the mother was satisfied.

Justice was served without punishing anyone. The delivery company did not sanction or fire its employee. In fact, the company was a bit angry initially because it said the hospital had never told them how it would like deliveries to be done. So why did they suddenly come with this now?

But without knowing it, the parties had asked and answered the restorative questions. Who got hurt? The baby, the mother. What were their needs? Reassurance that it won't happen again. Whose obligation was it to meet those needs? The hospital's, in consultation with the delivery contractors. And in developing the solution, the community was involved: relevant people in the hospital, the mother, the contractors.

Then, two years later, another delivery contractor knocked over a woman, resulting in a leg fracture. Having learned from its spontaneous restorative processes, the hospital and delivery company apologized, agreed to manage and cover the costs of treatment, and made some additional adjustments to its delivery standards and practices. Again, nobody was sanctioned or fired, and no punitive damages were paid. The woman felt that justice was served.

Let's look at the various steps needed to create restorative justice. This includes finding out who has been hurt, what the person(s) needs are, and whose obligation it is to meet those needs. Restorative practices are focused on keeping the "offender" in the community rather than separating or exiling him or her, and to have the community play a big role in the restorative practice.

Restorative Justice Steps

Who Was Hurt, and What Are His or Her Needs?

The first questions asked by restorative justice is who was hurt and what his or her needs are. An incident in a hospital, for instance, or an airline, oil company, or other organization can hurt various (groups of) people. Recognizing the ways in which they hurt and responding to their needs is necessary if you want your organizational culture to become truly just.

- **First victims**: Patients, passengers, colleagues, or surrounding community who suffer the consequences. Their needs might center on information about the incident, access to the practitioners involved, and some type of restitution and reassurance of prevention.

- **Second victims**: The practitioner(s) involved who feel(s) personally responsible and suffer(s) as a result. They might need anything from empathy and compassion, to opportunities to show remorse, to counseling and trauma care.
- Your **organizational community** also has needs. This is likely information about the incident and the organizational response to it. But they may also want an opportunity to help first and second victims, contribute to restoring relationships and trust, and achieve a sense of joint problem ownership.

What does each group need? First victims typically need information about the incident. What happened, and why? First victims are quick to see through speculation or legally constrained information. Access to the practitioner(s) involved can be an important way to help get them some of the authentic information they might crave. It is well known that lawsuits in the wake of patient harm are significantly less likely when there is immediate, open, and honest disclosure of what happened.[32] This suggests that many first victims need honesty and information and an acknowledgment of their humanity more than they want financial compensation.

The daughter of a woman who was injured after receiving a medication to which she had a documented allergy commented on her mother's preserved trust in her physician: "The reason [the physician's] apology felt genuine was because it was direct. He didn't beat around the bush. He didn't try to cover things up." Rather than simply assigning blame, patients and families want both to understand their situation fully and to know what the event has taught caregivers and their institutions.[33]

First victims also need an opportunity (or multiple opportunities) to tell *their* story. The incident likely disrupted their trust in the system or its practitioners. Telling and retelling their story can help them integrate the incident into their worldview, to give it a place with some boundaries around it so that it does not forever keep affecting everything they do and are.

First victims might also feel a need to regain control over their experiences and emotions. They are likely to feel upset, betrayed, angry, disillusioned, disappointed, and confused. One way to help them is empowerment, for instance by involving them in investigative and restorative justice processes. First victims can help define the kind of obligation that will be asked of (and agreed with) the "offender." This gives first victims an active part in determining proportionality, through which they can attempt to express what the incident meant to them—while keeping in mind the humanity of the second victim, the practitioner(s) involved. First victims also might need some sort of restitution. If their actual loss cannot be compensated, first victims typically want to know that everything is going to be done to prevent recurrence. They don't want others to suffer like they did. This might even lead to the first victim urging justice for the second victim, as in the case that follows.

Air traffic controllers in Yugoslavia were charged with murder and were jailed in the wake of a midair collision between two passenger aircraft. One hundred and seventy-six lives were lost. It was 1976, and Zagreb was one of the busiest air traffic control centers in Europe. Its navigation beacon formed a crossroads of airways heavily used by traffic to and from southeastern Europe, the Middle East, the Far East, and beyond. The center, however, had been structurally understaffed for years. At the time of the accident, the radar system was undergoing testing and the center's radio transmitters often failed to work properly. Through a combination of different languages and flawed data presentation to the controller, one of the aircraft managed to level off exactly at the altitude of another. Three seconds later, its left wing smashed through the other's cockpit and both aircraft plummeted to the ground. "Improper air traffic control operation," the accident investigation concluded. One controller, however, was singled out and sentenced to a prison term of seven years, despite officials from the aviation authority offering testimony that the Zagreb center was understaffed by at least 30 controllers. Significantly, the father of one of the victims of the collision led an unsuccessful campaign to prevent the controller's jailing. He then joined the efforts of other controllers to have him released after serving two years.[34] *One of the major reasons for his efforts was that he, as a first victim, did not believe that jailing the controller was fair to either the first or the second victims. First victims got no assurance of any improvements and possible prevention of repetition, and second victims (practitioners from the sharp end) were unfairly singled out for what was the failure of an entire complex system and the organizations set up to manage it. It was not until the early 1990s that the air traffic control system around Zagreb was revamped.*

What do second victims typically need? For most professionals, an error that leads to an incident or death is antithetical to their identities. They themselves can see it as a devastating failure to live up to their professional commitment. Having made an error in the execution of a job that involves error management and prevention is something that causes excessive stress, depression, anxiety and other psychological ill health. Particularly when the work involves considerable autonomy and presumptions of control over outcomes on the part of the actor (such as doctors, pilots, air traffic controllers), guilt and self-blame are common, with professionals often denying the role of the system or organization in the spawning of their error altogether and blaming themselves entirely. This sometimes includes hiding the error or its consequences from family and friends. Practitioners might distance themselves from any possible support, or attempt to make atonement themselves with those who were harmed by the error. The memory of error stays with professionals for many years.[35] All of these effects are visible, and can be present strongly even before your organization does anything, or before a manager or prosecutor might do anything.

In the best case, second victims seek to process and learn from the mistake, discussing details of their actions with colleagues or employers. Practitioners can punish themselves harshly in the wake of failure. You or your organization or society can hardly make such punishment any worse—other than confirming what the practitioner already feels. As told by one physician in 1984 about a couple whose pregnancy was lost:

> ...although I told them everything they wanted to know and described to them as completely as I could what had happened, I never shared with them the agony that I underwent trying to deal with the reality of events. I never did ask their forgiveness.... Somehow, I felt it was my responsibility to deal with my guilt alone.[36]

What second victims typically need is an opportunity to tell their story, to not feel alone, singled out, vilified, or shunned. They probably do want to offer their account of what happened; they too want to find out how things could go so wrong and what they and others could do differently in the future. They may need anything from empathy and compassion to counseling and trauma care. These are among the things your organization might consider:

- Second victims need to regain trust in their own competencies, and rebuild relationships with others who rely on them.
- They also want reassurance that they, their organization, and their community have put things in place to prevent recurrence.
- They may want to contribute to an investigation, and help by suggesting countermeasures or improvements. This gives them an opportunity to convert their feelings of guilt into social or practical action.

The role of the organization or peers in facilitating such coping is important. This can be done through peer or managerial support and appropriate structures and processes for learning from failure that might already be in place in your organization. Research on employee assistance programs stresses that employees must not get seen as the source of the problem, or treated as somehow "troubled" as opposed to "normal" employees. Social support, and particularly peer support, can then be a crucial moderator of the stress, anxiety, and depression that a second victim can experience in the aftermath of an incident. Such support is also a strong predictor of the second victim coming out psychologically healthy. Guidance on setting up effective peer support and stress management programs in the wake of incidents is available in separate work.[37] I have written much more about the experience of the second victim, and about trauma, guilt, resilience, and forgiveness in the book *Second Victim*.[9]

Criminalization, the topic of Chapter 4, makes things a lot worse for the second victim. Criminalization affirms feelings of guilt and self-blame and exacerbates their effects, which are linked to poor clinical outcomes in other settings.[38] It can lead to practitioners going on sick leave, divorcing, exiting the profession permanently, or even committing suicide. Another response, though much more rare, is an expression of anger and counterattack, for example through the filing of a defamation lawsuit. Criminalization can also have consequences for a person's livelihood (and his or her family), as licenses to practice may be revoked. This in turn can generate a whole new layer of anxiety and stress. One pharmacist, whose medication error ended in the death of two patients, suffered from depression and anxiety to such an extent that he eventually stabbed his wife to death and injured his daughter with a knife.[39]

Identifying the Obligations to Meet Needs

Wrongs or harms result in obligations. In a restorative just culture, these obligations are acknowledged and articulated. They get met by different stakeholders, preferably in collaboration with each other.

The practitioner (or second victim) can, for example, be obliged to

- Honestly disclose his or her role in the incident and give an account to the others involved or affected
- Recognize the needs of first victim(s), organization, and community
- Show remorse and be open to various ways to put things right with first victim, organization, and community
- Identify pathways to prevention in collaboration with first victim(s), organization, and community

The organization and surrounding community can embrace the obligation to

- Offer support to first and second victims (through open disclosure or critical incident stress management programs)
- Not fire or sanction people just because they were involved in an incident
- Ask itself honestly *what* was responsible for the incident, not *who*
- Perform an investigation on the premise that people did not come to work to do a bad job, and one that asks why it made sense for people to do what they did
- Identify pathways to prevention, in collaboration with first and second victim(s)

The first victim typically has an obligation to

- Respect the humanity of the practitioner involved in the incident
- Be willing to be part of the solution, for example, by contributing ideas for possible prevention

I learned of an incident in an airline not long ago, to which the chief executive and safety manager had very different responses. On the back of significant expansion, the airline had hired many new cabin crew members. The ones in the back galley were typically the junior ones and had to work really hard on short flights (with short turnaround times at each end) to get everything done that they needed to do. Right after landing, a junior crew member in the back galley noticed that a catering cart slid out of its place and moved forward when the aircraft started braking after landing. Strapped into her jumpseat, she tried to grab it, but it was too late. The cart careened down the aisle, all the way to the front of the aircraft. On the way, it crushed a passenger's foot and then caused significant damage in the front of the cabin where it came to rest.

The CEO of the airline, on hearing about the incident, decided to write an angry email message to all of his managers, impressing upon them the need to make sure their people followed the rules to the letter and were careful and vigilant in their work. A catering cart had not been properly secured! How could that even happen if people were simply doing the right thing? Stuff was going to get broken this way, and it was going to cost the airline a lot of money.

The safety manager had a very different response. Without knowing anything about systems of restorative justice, he asked the very questions that animate it. Who was hurt? What were the person(s) needs? Whose obligation was it to meet those needs? The passenger was clearly hurt. He needed medical care. He needed bills paid. He needed an apology, or various apologies. But the cabin crew member was hurt too. She was devastated about what had happened, and fearful of losing her job. Other passengers too, as a community, may have been frightful of what they saw happen. Trust in the airline might have been hurt. Carefully plotting his way through the tense political landscape in the aftermath of this incident, the safety manager was able to get people to talk to each other about the hurts, needs, and obligations that had resulted from it. He was successful. The passenger was satisfied, the cabin crew member kept her job, and even the CEO was content with the outcome.

Identifying and meeting obligations is ultimately about putting right what went wrong. It is about making amends. In restorative practices, this means promoting reparation and healing for all affected by the incident. This notion of reparation, of restitution or "paying back," is central to retributive justice too, of course. In restorative practices, however, everything possible is done to reintegrate the practitioner into the community, and the "payment" typically is made in a different currency. Restorative practices ask you to

- Address the harm done to first and second victims of the incident, as well as the surrounding community
- Address the systemic issues that helped produce the incident by asking *what* was responsible for it, so that other practitioners and first victims are less likely to end up in a similar situation

For restorative practices to be meaningful and seen as just by all involved, you have to be collaborative and inclusive. Effective restoration relies on this engagement. An incident can affect many people, and its aftermath typically has many stakeholders. These might be given access to, and information about, each other. All can then be involved in deciding what justice requires in their case. This may mean an actual dialogue between parties (e.g., first and second victim), to share their accounts and arrive at an agreement on what should be done. How might the creation of restorative justice look in your organization? It will likely involve the following steps and people:

- Encounters between stakeholders. The first one is likely to be between your organization and the practitioner(s) involved in the incident. Remember your organization's obligations above!
- An encounter between first and second victims, appropriately guided, may follow. Surrogates or representatives may need to be used in some situations.
- Encourage all stakeholders to give their accounts, ask questions, express feelings, and work toward a mutually acceptable solution.
- Acknowledge the harm, restore the balance, and address your future intentions.

To succeed, you will need to broaden out the conversation about a restorative just culture. Where necessary, include senior management, the board, your regulators, human resources, your safety department, unions or professional associations, customers, and other stakeholders. The more a solution is created and supported by the community from which both second and first victims stem, the more likely it is that such a solution is accepted by them and acted upon. This even goes for any sanctions you might want to come up with. Indeed, research shows that the severity of sanctions is a poor predictor of the effectiveness of the sort of social or managerial control you wish to exercise through them. Instead, the extent to which such sanctions are socially embedded in the community is a stronger predictor.[31] In other words, if you want your practitioners to act differently, do not scare them by separating out one of them and turning him or her into an "example." Rather, engage them with the community that will have to keep working together to ensure safe work, to ensure future successful outcomes. There is nothing wrong with having a practitioner feel temporarily out of favor—if indeed pathways for restoration are offered (see later under forgiveness). But your organization will gain nothing from a climate in which practitioners are constantly fearful and insecure in their relationship to you and to each other. That surely is a culture without trust. And such a culture can never be a just culture.

Ask yourself these questions to check how close to restorative just culture you and your organization might be:

1. Does your just culture process address harms, needs, and causes?
2. Is it adequately victim oriented (including both first and second victims)?
3. Are practitioners encouraged to recognize their contribution to the (potential) harm caused, but also treated as potential second victims?
4. Are all relevant stakeholders involved?
5. Is it based on dialogue, participation, and collaborative decision making?
6. Does it identify and address deeper, systemic issues that gave rise to the incident in the first place?
7. Is it respectful to all parties?

Restoration and Forgiveness

There is of course a link between restorative practices and forgiveness. But forgiveness is not necessarily the goal of restorative justice. It can be one of the outcomes, but one party can never be obligated to forgive another—that would be a meaningless, hollow act. Forgiveness is fundamentally relational.[17] Forgiveness, as well as the processes of disclosure (or confession) and apology (or repentance) that necessarily precede it, has religious connotations. But that is no reason to dismiss them as irrelevant to a just culture in your organization. In fact, Nancy Berlinger, a colleague at the Hastings Center in New York, suggests that there is a lot that you can do in your organization to put conditions in place that make disclosure, apology, and forgiveness possible, often at little or no monetary cost. Such conditions include

- Promptly acknowledging an error and offering the first victim(s) an authentic account of what happened.
- Being sympathetic to calls for accountability even when the system has contributed considerably to producing the failure. Some first victims may have trouble understanding system failure, seeing only personal shortcomings. Others may be keen to see the organization move beyond its focus on the individual and address system-level issues instead.
- Providing opportunities for second victims to process incidents and receive support in an environment that is neither punitive nor demeaning.
- Nurturing a commitment that withholding the truth violates the humanity and autonomy of the first victim, and has a corrupting effect on second victims and their colleagues.
- Avoiding the sort of scapegoating of subordinates that would diminish your own responsibility.
- Avoiding assertions that the first victim was somehow to blame (e.g., non-compliant patient, obese passenger who could not evacuate quickly enough, wayward pilot who did not follow air traffic control instructions).[40]

In one case, the first victim actually made public disclosure a condition for forgiveness. For her to desist from legal action, the second victim had to write about the surgical error he had made and publish it in a top journal in the field. He did, and it appeared in Surgical Endoscopy *in 1995. The first victim kept her word, and the second received many positive responses.*[41]

Disclosure is incomplete without apology. And both are preconditions (though not guarantees) for receiving forgiveness. Apology (or, more correctly, repentance) means allowing yourself to feel, and express, sincere sorrow and regret for what you have done, and for what has happened as a result. Both disclosure and apology contribute to accountability: the giving of an account that includes narrative, explanation, expressions of regret, and even admissions of guilt and responsibility.

Disclosure and apology may still be an act of healing in cases where the relationship between the first and second victims itself is harder to heal. Writing in the New England Journal of Medicine, *a surgeon told of performing a wrong operation on a 65-year-old woman's left hand, something he managed to correct not long after.*[42] *Yet in this case, a relationship with the first victim was no longer really possible. The patient's son received apologies, fee waivers, and offers of follow-up care for the mother. But she had lost faith in the doctor and would not return to have her sutures removed, or receive any other care, or ask for an apology or explanation. There was nothing the surgeon could do to demand contact or urge forgiveness.*

Organizational practices that could support repentance include

- Not forcing the first victim to interact with the second (i.e., the practitioner involved in their injury or loss) if there is no wish to do so—and vice versa

- Appreciating the difference between appropriate feelings of guilt ("I *made* a mistake") and destructive feelings of shame ("I *am* a mistake") that might be felt by second victims
- Offering first victims and their families access to care or other services should they need them
- Meeting the obligations that result from the incident, injuries, or loss on the part of the first victim(s)
- Recognizing that asking a first victim for "forgiveness" may be obtrusive or culturally inappropriate, while at the same time working to create conditions that may allow both victims, in their own time, to detach from the incident as a continuing source of pain, anger, and injustice[40]

Organizational practices can come very close to forgiveness, or at least create appropriate conditions, without demanding it. The relational and restorative nature of these practices can easily be recognized:

- Inviting first and second victims to be part of your organization's or industry's quality and safety improvement processes, though not making it their responsibility
- Offering safe places or rituals for second victims to explore their reactions and responsibilities concerning errors and incidents
- Identifying and changing aspects of the professional culture in your organization that deny the fallibility, and therefore the humanity, of your practitioners
- Identifying and changing features of your organization that work against truth-telling, accountability, compassion, and justice in dealing with incidents[40]

COMPARING AND CONTRASTING RETRIBUTIVE AND RESTORATIVE APPROACHES

Remember the different sets of questions asked by retributive versus restorative approaches to justice (see Table 1.1). Notice the different ways of thinking about accountability in retributive versus restorative just cultures:

- In retributive justice, an account is something you *pay*. Retributive justice holds people accountable by making them repay the debt they morally owe. People have to settle their account with their organization, victims, community, and society.
- In restorative justice, an account is something you *tell*. In restorative justice, people together figure out how to offset the harm. It holds people accountable by having different parties tell their often contradictory and only partially overlapping accounts.

The line between retribution and restoration is not entirely clear. You could actually do some of both, at the same time. Both retribution and restoration can be seen

TABLE 1.1
Contrasting Retributive and Restorative Questions to Ask

Retributive	Restorative
Which rule has been broken?	Who has been hurt?
Who did it?	What are the person(s) needs?
How bad is the infraction, and so what does the person deserve?	Whose obligation is it to meet those needs?

or treated as processes that help reintegrate a person into a community. In some cases, retribution might even be a precondition for restoration. As in: *First pay up as a sign that you are taking responsibility. Once that is settled, we'll happily have you back—even if you are a bit poorer, or with less status.* I have seen pilots who had been demoted in the wake of an incident in the crew room, behaving as if nothing had happened. But something had happened, and they had one fewer stripe on their shoulders to show for it. Yet there they were, happily laughing and chatting away with their colleagues, part of the "tribe" just like before.

So we shouldn't overstate the contrast between retribution and restoration. For instance, whether it is done under regimes of restoration or retribution, some form of "penance" is fundamental to processes of forgiveness. This involves telling your story, expressing that you are sorry, and in some way accepting (part of the) responsibility for the outcome. Of course, retributive and restorative practices have different ways of getting people to tell that story and accepting responsibility. And they may get different stories as a result. However,

- Both retributive and restorative forms of justice and accountability acknowledge that a "balance" has been thrown off by the act and its consequences.
- Both also understand that there needs to be a proportional relationship between the act, the consequences, and the response to it.
- Thus, both are designed around some sense of reciprocity, of "evening the score." They differ, however, on the "currency" that is used to rebalance the situation, to even that score, to fulfill the obligations.[43]

Neither Retributive nor Restorative Justice "Lets People Off the Hook"

Neither form of just culture gets "people off the hook." Both hold people accountable. In both, people are expected to engage with, and respond to, the community of which they are, or were supposed to be, part. Both forms of just culture impose accountability. But they go about it in different ways. Retributive justice achieves accountability by looking *back* on the harm done, or potentially done, by the person.

- It asks what the person must do to compensate for his or her actions or consequences.
- Justice is created by meeting hurt with hurt. (Potential) hurt is compensated by imposing more hurt—deserved and proportional.

- People can feel that the person is held accountable by not letting him or her off the hook.
- The community can demonstrate that it does not accept what the person did (it would not accept such actions from *any* of its members) and demonstrates that it makes the person pay.

The focus in restoration, in contrast, is not chiefly on what some specific "offender" deserves, like it is in retributive justice. Restorative justice achieves accountability by looking more *ahead* at what must be done to repair the trust and relationships that were harmed by the person's actions.

- This makes it important for others to understand why it made sense for the person to do what he or she did, and how others could perhaps be put in the same situation.
- Rather than seeing the "offenders" as causes of trouble, restorative practice will tend to see "offenders" as inheritors of organizational, operational, or design issues that could set up others for failure as well.
- Restorative practices are thus likely to get to the systemic issues that triggered the incident, to identify the deeper conditions that allowed an incident to happen.
- For this to work, of course, the people involved need to tell their account, their story. This also gives them the opportunity to express remorse for what they did or for what happened, should that be appropriate.
- The people affected by the incident, as well as other community members, can explore and agree on what needs to be done to restore trust and relationships.
- The community demonstrates that it expects people to be accountable by getting them to reflect on their behavior and sharing the insights.

Retributive and Restorative Forms of Justice Deal Differently with Trust

The two forms of just culture also approach trust differently. Retribution builds trust by reinforcing rules and the authority of certain parties or persons to police and enforce them. It says that where people work to get things done, there are lines that should not be crossed. And if they are, there are consequences. Think about it like this: if you find that people "get away" with breaking rules or doing sloppy work, you don't have much trust in the system, or in your community's ability to demand accountability. Your trust can be restored if you see an appropriate and assertive response to such behavior. You can once again rest assured that the system, or your community, does not accept such behavior and responds in ways that make that clear—to everyone.

Restoration, on the other hand, builds trust by repairing fiduciary relationships. Fiduciary relationships are relationships of trust between people who depend on each other to make something work. Consider the work done in your own organization. People in your organization depend on each other. Every day, perhaps every minute, they have to trust each other that certain things get done, and get done in a timely, appropriate, and safe manner. They might not do these things themselves because

TABLE 1.2
The Different Ways in Which Retributive and Restorative Processes Try to Create Justice

Retributive	Restorative
Believes that wrongdoing creates guilt, and demands punishment that compensates it	Believes wrongdoing creates needs, and obligations to meet those needs
Believes an account is something the offender *pays* or *settles*	Believes an account is something the offender *tells* and listens to
Asks *who* is responsible for the incident	Asks *what* is responsible for the incident
Learns and prevents by setting an example	Learns and prevents by asking why it made sense for people to do what they did
Focuses on what people involved in the incident deserve	Focuses on what people involved in, and affected by, the incident need
Creates justice by imposing proportional and deserved punishment	Creates justice by deciding who meets the needs arising from the incident
Meets hurt with more hurt	Meets hurt with healing
Looks back on harm done, and assigns consequences	Looks ahead at trust to repair, and invests in relationships
Builds trust by reinforcing rules and the authority to impose and police them	Builds trust by repairing relationships between people who depend on each other to accomplish their work

they are not in the right place, or because they lack the expertise or authority to do them. So they depend on others. This creates a fiduciary relationship: a relationship of trust. It is this relationship that is hurt or broken when things go wrong. And it is this relationship that needs restoring.

Both kinds of trust can be important for your organization or community (see Table 1.2).

CAN SOMEONE OR SOMETHING BE BEYOND RESTORATIVE JUSTICE?

Retributive theory believes that pain will vindicate.[43] That is, responding to hurt with more hurt will somehow equalize or even eliminate the injustice that has been inflicted. Restorative theory, instead, believes that pain requires healing. But are there cases where those who have inflicted the pain are beyond healing, beyond the reach of restorative justice? Advocates of restorative approaches might like to believe that nothing or nobody is beyond the reach of restoration or reintegration. Yet many others can point to cases in which they feel a retributive response is the only appropriate one. Some cases may call for a process that gives attention to societal needs and obligations above all others—particularly above the needs of any immediate stakeholders (e.g., first and second victims).

It isn't the case itself that determines whether it is beyond restoration. Rather, it is our judgment about the case; it is about what we find important, what we find "just" or the morally right thing to do. So always ask the question: *who* gets to decide whether a case is beyond the reach of restorative approaches, and what are their stakes (if any) in saying it is so?

If the decision to forego restoration is made and generally agreed with, there are a few of things to remember.

- Retributive justice is often criticized for not being sufficiently victim oriented. First-victim oriented, that is. You will find some examples of this in Chapter 4. First victims may feel left out or sidelined. They have a stake in the creation of justice, but are often given no voice to contribute toward this end.
- Retributive justice is also criticized for not involving the community enough, for not embedding sanctions socially, for doing its own thing essentially between two parties (the party doing the judging and the party being judged). Retributive justice engages these parties in a process that is mostly removed from the rest of the community, often conducted in a language that is alien to that used in practice in the community, and away from the time and the place where the incident happened.
- Openness to different accounts of what happened can get sacrificed in an adversarial setting where one account wins and one account loses. As a result, not much of value might be learned; not many systemic improvements may follow from retributive justice.
- Retributive approaches can encourage "offenders" to look out for themselves and discourage them from acknowledging their responsibility in any concrete ways because it might self-incriminate them even further.

If you can find ways to mitigate these negative aspects of retributive justice, it will likely help others see your responses as more "just." So whatever you do, ask who is hurt. Give a voice to the different stakeholders. Identify responsibilities and obligations that various parties need to meet—not just the "offender" or second victim. Try to socially embed your responses, so that the community feels part of the solution.

CASE STUDY

Are All Mistakes Equal?

Not all rule-breaking can be seen the same way. Or can it? Some actions of your people must be considered to be worse than others. Not all breaches of trust are the same. Not all actions, or even mistakes might be equally "forgivable." Let's look at one basic distinction that can be made in many professions: that between technical and normative errors. These are not types of errors that exist "out there" in the world. Rather, they are ways of constructing, or looking at and talking about, your people's actions. Sometimes professions themselves have ways of making such distinctions, because they do useful work for them, their colleagues, their training and selection, and their organizations. Again, it does not mean that these categories are ready formed and should be hunted down and exposed. Instead, it is a way of looking at the way in which other people talk about error, and how they might end up with a judgment of whether the error is forgivable or not, whether the mistake is less equal than others.

When studying the way surgeons treat errors that can hurt (or have hurt) patients, Charles Bosk, a sociologist, saw a remarkable pattern. Surgeons and other physicians made a distinction between what he began to call *technical* and *normative* errors.[44] To be sure, it was not the error that is either technical or normative. It became technical or normative because of the way people looked at the error, because of what they saw in it, talked about it, and how they responded to it. The distinction can have powerful consequences for how your organization (or surrounding society) is prepared to deal with an error that occurred. Whether you construct an error as normative or technical has far-reaching consequences for exacting accountability and encouraging learning.

After this, we need to consider yet another really important factor in our judgment of whether a mistake is forgivable or not: knowing the outcome. Hindsight plays a huge role in how we handle the aftermath of a mistake. We assume that if an outcome is good, then the decisions leading up to it must have been good too—people did a good job. And it works in the opposite too. If the outcome of a mistake is really bad, or could have been really bad, we are more likely to see the mistake as culpable. There is more to account for. I will present a case that shows that knowing how things turned out determines whether we see actions as culpable or not.

Technical Errors: Errors in a Role

When a practitioner makes a technical mistake, she or he is performing her or his role diligently, but the present skills fall short of what the task requires.

For example, when a pilot makes a hard landing, this could likely be the effect of skills that have yet to be developed or refined. Technical errors will be seen as opportunities for instructors or colleagues to pass on "tricks of the trade" (e.g., "start shifting your gaze ahead when flaring").

People can be quite forgiving of even serious lapses in technique, as they see these as a natural by-product of learning-by-doing. Technical errors do not just have to be connected to the physical handling of a process or its systems; they can also involve interaction with others in the system, for example, air traffic control, nurses, night staff, physicians. The person in question, for example, may have seen the need to coordinate (and may even be doing just that), but does not have the experience or finely developed skills to recognize how to be sensitive to the constraints or opportunities of the other members of the system.

For an error to be constructed as technical, however, it has to meet two conditions:

- One is obviously that the frequency or seriousness should decrease as a person gains more experience. When someone keeps making the same mistakes over and over, it may be difficult to keep seeing them as purely technical. However, as long as the person making the errors shows a dedication to learning and to his or her part in creating safety, that person is still conscientiously filling his or her role.
- The other condition for a technical error is that it should not be denied, by the pilot involved, as an opportunity for learning and improvement. If a practitioner is not prepared to align discrepant outcomes and expectations by looking at him- or herself, but rather turns onto the one who revealed the discrepancy, trainers or supervisor or managers (or courts) will no longer see the error as purely technical.

A flight instructor reports: We had been cleared inbound to a diversion airport due to weather. We were on downwind when an airliner came on the frequency and was cleared for the instrument landing system (ILS) toward the opposite runway. The student proceeded to extend his downwind to the entry point he had chosen, even though the field was now fully visible. He was entirely oblivious to hints from air traffic control to turn us onto the base so we could make it in before the airliner from the opposite side. When the student still did not respond, I took control and steered our aircraft onto the base. We completed the landing without incident before the airliner came in. On debriefing, however, the student berated me for taking control, and refused to accept the event as an opportunity to learn about "fitting in" with other traffic at a dynamic, busy airport. He felt violated that I had taken control.

Professionals with limited experience may not be very sensitive to the unfolding context in which they work. They have simply not yet learned which cues to pick up on and how to respond to them. People will see such insensitivity as a technical issue consistent with the role of student, and a due opening for enlightenment. Sticking to the plan, or behaving strictly in the box, even though a situation has unfolded differently, has been known to lead to problems and even accidents. So valuable lessons are those that demonstrate how textbook principles or dogged elegance sometimes have to be compromised to accommodate a changing array of goals. Professionals can otherwise end up in a corner. Surgery has a corollary here: "Excellent surgery makes dead patients."

The benefits of technical errors almost always outweigh the benefits. Of course, this is so in part because the division of labor between senior and junior practitioners in most operating worlds (or between instructors and students) is staggered so that no one advances to more complex tasks until he or she has demonstrated his or her proficiency at basic ones.

Bosk tells how Carl, a surgical intern, was closing an incision, while Mark, the chief resident, was assisting. Carl was ill at ease. He turned to Mark and said "I can't do it." Mark said, "What do you mean, you can't? Don't ever say you can't. Of course you can." "No, I just can't seem to get it right." Carl had been forced to put in and remove stitches a number of times, unable to draw the skin closed with the proper tension. Mark replied, "Really, there is nothing to it," and, taking Carl's hand in his own, said, "The trick is to keep the needle at this angle and put the stitch through like this," all the while leading Carl through the task. "Now, go on." Mark then let Carl struggle through the rest of the closure on his own.

If aid is necessary, there are almost always only two responses:

- Verbal guidance is offered, with hints and pointers.
- Or the superordinate takes over altogether.

The latter option is taken when time constraints demand quick performance, or when the task turns out to be more complex than people initially assumed. This division of labor can also mean that subordinates feel held back, with not enough opportunity to exercise their own technical judgment. The preceding example could be an instance of this, where the division of labor is seen by the student as stacked in favor of the instructor. For instructors, supervisors, managers, and others, the challenge is always to judge whether the learning return from letting the practitioner make the mistake is larger than from helping her or him avert it and clearly demonstrating how to do so.

In another example, Bosk tells of the difficulty of performing a myelogram (a diagnostic procedure involving the removal of spinal fluid and the injection of dye in the spinal column) that had been ordered for a patient named Mr. Eckhardt. A senior student was to instruct a junior student in the procedure. They tried without any success to get the needle in the proper space. After some fumbling and a few sticks at Eckhardt, the senior student instructed the junior student to go "get Paul" (a second-year resident). Paul came in and surveyed the situation. After examining Eckhardt's back he told the students, who were profusely apologizing for their failure, not to worry; that the problem was in Mr. Eckhardt's anatomy and not in their skills. He then proceeded with some difficulty to complete the procedure, instructing the students all the while.

As for professionals, they should not be afraid to make mistakes. They should be afraid of not learning from the ones that they do make. Bosk's study showed how self-criticism is strongly encouraged and expected of surgeons in the learning role (which is to say, almost every surgeon). Everybody can make mistakes, and they can generally be managed.

Denial or defensive posturing instead discourages such learning. It allows the trainee or subordinate to delegitimize mistake by turning it into something shameful that should be brushed aside, or into something irrelevant that should be ignored.

Denying that a technical error has occurred is not only inconsistent with the idea that they are the inevitable by-product of training; it also truncates an opportunity for learning. Work that gets learned-by-doing lives by this pact: technical errors and their consequences are to be acknowledged and transformed into an occasion for positive experience, learning, and improvement. To not go along with that implicit pact is no longer a technical error; it is a normative one.

Normative Errors: Errors of a Role

Technical errors say something about the professional's level of training or experience. Normative errors say something about the professional him- or herself relative to the profession. Normative errors are about professionals not discharging their role obligations diligently.

- Technical errors create extra work, both for superordinate and subordinate. That, however, is seen as legitimate: it is part of the trade, the inevitable part of learning by doing, of continuous improvement.
- The extra work of normative errors, however, is considered unnecessary.

In some cases, it shows up when a crewmember asserts more than his or her role allows:

> A senior airline captain told me about one case that he constructed as a normative error. It was my turn to go rest, he said, and, as I always do, I told the first and second officers, "If anything happens, I want to know about it. Don't act on your own, don't try to be a hero. Just freeze the situation and call me. Even if it's in the middle of my break, and I'm asleep, call me. Most likely I'll tell you it's nothing and I'll go right back to sleep. I may even forget you called. But call me." When I came back from my break, it turned out that a mechanical problem had developed. The first officer, in my seat, was quite comfortable that he had handled the situation well. But I was a bit upset. Why hadn't he called me? How can I trust him next time? I am ultimately responsible, so I have to know what's going on.

The situation was left less resilient than it could (and, in the eyes of the captain, should) have been: leaving only two more junior crewmembers, with no formal responsibility, in charge of managing a developing problem. Of course, there are potential losses associated with calling:

- The superordinate could think the call was superfluous and foolish, and get cranky because of it (which the first officer in this example may have expected and, as it turned out, misjudged).
- The subordinate foregoes the learning opportunity and gratification of solving a problem her- or himself.

But the safe option when in doubt is always to call, despite the pressures not to. That is, in many cases, how a subordinate crewmember is expected to discharge her or his role obligations. In other cases, fulfilling those obligations is possible *only* by breaking out of the subordinate role, as a chief pilot once told me:

> My problem is with first officers who do not take over when the situation calls for it. Why do we have so many unstabilized approaches to runways in (a particular area of our network)? If the captain is flying, first officers should first point out to him or her that he or she is out of bounds, and if that does not work, they should take over. Why don't they, what makes it so difficult?

The chief pilot here flagged the absence of what may turn out critical for the creation of safety in complex systems: the breaking-out of roles and power structures that were formally designed into the system. Roles and power structures often go hand-in-glove (e.g., captain and first officer, doctor and nurse), and various programs (e.g., crew resource management training in aviation and healthcare, team resource management in air traffic control, bridge resource management in shipping, and so forth) aim to soften role boundaries and flatten hierarchies. These programs want to increase opportunities for coordinating viewpoints and sharing information. Where people do not do this, they fail to discharge their role obligations too—in this case by not acknowledging and deploying the flexibility inherent in any role.

Perhaps the fact that others see this as a normative breach is not so strange. In the kinds of operating worlds where we believe a just culture is important, it is very difficult to know and anticipate all the problems that may occur during a lifetime of practice. There will always be things that practitioners remain inexperienced with, simply because that kind of problem, in that kind of way, has not appeared before. Indeed, in complex and dynamic work, where resource limitations and uncertainty reign, failure is going to be a lasting statistical reality.

The possibility of suffering technical errors will consequently never go away entirely. In such worlds, where the knowledge base on how to create safety is inherently and permanently incomplete, many believe firmly in the importance of disclosing, discussing, and learning from error. When that does not happen, even an honest, technical error can become seen as a dishonest normative one.

"Covering up is never really excusable," Bosk quotes an attending physician as saying. You have to remember that each time a resident hides information, he is affecting someone's life. Now in this business it takes a lot of self-confidence, a lot of maturity, to admit errors. But that's not the issue. No mistakes are minor. All have a mortality and a morbidity. Say I have a patient who comes back from the operating room and he doesn't urinate. And say my intern doesn't notice or decides it's nothing serious and doesn't catheterize the guy and doesn't tell me. Well, this guy's bladder fills up. There's a foreign body and foreign bodies can cause infections; infection can become sepsis; sepsis can cause death. So the intern's mistake here can cost this guy hundreds of dollars in extra hospitalization and possibly his life. All mistakes have costs attached to them. Now a certain amount is inevitable. But it is the obligation of everyone involved in patient care to minimize mistakes. The way to do that is by full and total disclosure.[44]

The obligation to report or disclose, discuss, and learn seems to be a critical hinge in how we believe a just culture should work. But honest and open accounting can seem dangerous to many practitioners. How an error might be interpreted after-the-fact is sometimes entirely up for grabs. A technical one (missing an approach, or supplying the wrong drug because of inexperience with that particular drug or procedure or kind of patient) can easily be converted into a normative error—with much more serious consequences for accountability (such as a criminal trial).

[illegible]

2 Why Do Your People Break the Rules?

Why do your people break the rules? This is of course a normative question. It assumes that the rules are right to begin with. So perhaps it is not even the right question to ask! Be that as it may, you may actually have your own answers to this question. Is it because your people can't be bothered following them? Or have your people learned that those particular rules don't matter, or that a shortcut gets them to the goal faster without—supposedly—any harmful consequences? Or do they believe that they are above the rules; that the rules are really made for other people who are not as good at the job as they are?

The way you think about this matters. It determines how you respond to the people involved in an incident. It determines how you believe you should hold them accountable for their behavior. So before you set your mind to any particular explanation, have a look at the research that has been done into this. This research offers a number of possible reasons why your people break the rules. Some of these might overlap with your experiences and intuitions. The advantage of theories based on research is that the investigators have been forced to make their ideas and assumptions explicit. And, of course, they have had to test them. Does the theory actually hold up in practice? Does it correctly predict or explain the reasons for why people do what they do? Well, ask yourself what your experience and intuition tells you. This chapter takes you through the following:

- **Labeling theory**: Your people break the rules because you label their actions that way.
- **Control theory**: Your people break the rules because they get away with it.
- **Learning theory**: Your people break the rules because they have learned that a shortcut doesn't harm anything or anyone.
- **Subculture theory**: Your people break the rules because they see the rules as stupid or only there for lesser mortals.
- **The bad apple theory**: Some of your people break the rules because they are bad, evil, or incompetent practitioners.
- **Resilience theory**: Your people break the rules because the world is more complex and dynamic than the rules. The way to get work done is to keep learning and adapting.

These different explanations, as you will see, interact and overlap in various ways. Goal conflicts, getting the job done, learning how to not get in trouble when "breaking rules," for example, seem to play an important role in almost all of them. This also means that you can apply multiple explanations at the same time, to see them as different ways of mapping the same landscape.

LABELING THEORY

Labeling theory says your people break the rules because you label their behavior that way; you call it that. This is as simple as it is counterintuitive. Because isn't rule breaking a *real* thing, something that you can establish objectively? Not quite, it turns out.

Consider an air traffic controller who gives an aircraft a clearance to descend earlier than the agreement between two neighboring air traffic sectors allows. In her world, she is not breaking a rule: she is taking into consideration the fact that in this aircraft type it is really hard to slow down and go down at the same time, that her neighboring sector is very busy given the time of day and peak traffic, that they might put the aircraft into a holding pattern if it shows up too soon. She is, in other words, anticipating, adapting, or doing her job, and expertly so. She is doing what all users (and managers) of the system want her to do.

Or consider the nurse who works in a hospital where his ward has adopted a so-called red rule that tells him to double-check certain medication administrations with a colleague before administering the drug to a patient. Suppose he has given this very medication, in this very dose, to the very same patient that morning already, after indeed double-checking it. Now it is evening, his colleague is on the other side of the ward attending to a patient with immediate needs, and because of manpower reductions in the hospital, no other nurse is to be found. The patient requires the prescribed drug at a certain interval, and cannot really wait any longer. Whether the nurse breaks a rule or is doing what needs doing to get the job done (the job everyone wants him to do) is not so objectively clear. It is often only after things go wrong that we turn such an action into a big issue, into a broken rule for which we believe we need to hold someone accountable.

The formal name for this is labeling, or labeling theory. It says that rule breaking doesn't come from the action we call by that name, but from *us* calling the action by that name. Rule breaking arises out of *our* ways of seeing and putting things. The theory maintains that what ends up being labeled as rule breaking is not inherent in the act or the person. It is constructed (or "constituted," as Scandinavian criminological researcher Nils Christie put it) by those who call the act by that label:

> The world comes to us as we constitute it. Crime is thus a product of cultural, social and mental processes. For all acts, including those seen as unwanted, there are dozens of possible alternatives to their understanding: bad, mad, evil, misplaced honour, youth bravado, political heroism—or crime. The same acts can thus be met within several parallel systems as judicial, psychiatric, pedagogical, theological.[45]

Violations Seen from This Bench Are Just Your Imagination

I used to fly out of a little airport in the US Midwest. Between two hangars there was a wooden bench, where old geezers used to sit, watching airplanes and shooting the breeze. When nobody was sitting on the bench, you could see that it said, painted across the back rest: "Federal aviation regulation violations seen from this bench are just your imagination."

Whereas cases of criminalizing human error show that we sometimes drag rules and laws over an area where they have no business whatsoever, the bench at this little airfield showed a more hopeful side of the negotiation, or construction, of deviance. Even if it was meant tongue-in-cheek (though when I, on occasion, sat on that bench, I did see interesting maneuvers in the sky and on the ground) we, as humans, have the capacity to see certain things in certain ways. We have the capacity to distance ourselves from what we see and label things one way or another, and even know that we do so. That also means we can choose to label it something different, to not see it as deviance—when we are sitting on that bench, for example.

But, you might protest, doesn't rule breaking make up some essence behind a number of possible descriptions of an action, especially if that action has a bad outcome? This is, you might believe, what an investigation, a hearing, or a trial would be good for: it will expose Christie's "psychiatric, pedagogical, theological" explanations (I had failure anxiety! I wasn't trained enough! It was the Lord's will!) and show that they are all patently false. You might believe that if only you look at things objectively, and use reason and rationality, you can strip away the noise, the decoys, and the excuses and arrive at the essential story: whether a rule was broken or not. And if it was, then there might, or should, be negative consequences.

Keith Ramstead was a British cardiothoracic surgeon who moved to New Zealand. There, several patients died during or immediately after his operations, and he was charged with manslaughter in three of the cases. Not long before, a professional college had pointed to serious deficiencies in the surgeon's work and found that seven of his cases had been managed incompetently. The report found its way to the police, which subsequently investigated the cases. This in turn led to the criminal prosecution against Ramstead.

To charge professionals like Keith Ramstead with a crime is just one possible response to failure, one way of labeling the act and thus dealing with it. It is one possible interpretation of what went wrong and what should be done about it. Other labels are possible, too, and not necessarily less valid:

- *For example, one could see the three patients dying as an issue of cross-national transition: Are procedures for doctors moving to Australia or New Zealand and integrating them in local practice adequate?*
- *How are any cultural implications of practicing there systematically managed or monitored, if at all?*
- *We could see these deaths as a problem of access control to the profession: Do different countries have different standards for who they would want as a surgeon, and who controls access, and how?*
- *It could also be seen as a problem of training or proficiency-checking: Do surgeons submit to regular and systematic follow-up of critical skills, such as professional pilots do in a proficiency check every six months?*

- *We could also see it as an organizational problem: There was a lack of quality control procedures at the hospital, and Ramstead testified having no regular junior staff to help with operations, but was made to work with only medical students instead.*
- *Finally, we could interpret the problem as sociopolitical: What forces are behind the assignment of resources and oversight in care facilities outside the capital?*

It may well be possible to write a compelling argument for each of these explanations of failure, for each of these different labels. Each has a different repertoire of interpretations and countermeasures following from it. A crime gets punished away. Access and proficiency issues get controlled away. Training problems get educated away. Organizational issues get managed away. Political problems get elected or lobbied away.

The point is not that one interpretation is right and all the others wrong. The point is that multiple overlapping interpretations of the same act are always possible (and may even be necessary to capture its full complexity!). And all interpretations have different ramifications for what people and organizations think they should do to prevent recurrence.

Some interpretations, however, also have significant negative consequences for safety. They can eclipse or overshadow all other possible interpretations. The criminalization of human error seems to be doing exactly that. It creates many negative side effects, while blotting out other possible ways forward. This is unfortunate and ultimately unnecessary.

But as the preceding examples show, the same action can be several things at the same time. It depends on what questions you asked to begin with. Ask theological questions—as we have done for a long time when it comes to rule breaking—and you may see in it the manifestation of evil, or the weakness of the flesh. Ask psychological questions and you might see smart human adaptations, expertise, and the stuff that is all in a day's work. Ask pedagogical questions and you may see in it somebody's underdeveloped skills. Ask judicial or normative questions and you may see a broken rule. Actions, says labeling theory, do not contain rule breaking as their essence. We make it so, through the perspective we take, the questions we ask. As Becker, an important thinker in labeling theory, put it:

> …deviance is created by society…social groups create deviance by making the rules whose infraction constitutes deviance and by applying those rules to particular persons and labeling them as outsiders. From this point of view, deviance is not a quality of the act the person commits, but rather a consequence of the application by others of rules and sanctions to an 'offender'. The deviant is the one to whom the label has successfully been applied; deviant behaviour is behaviour that people so label.[46]

Labeling theory, then, says that what counts as deviant or rule breaking is the result of processes of social construction. If a manager or an organization decides that a certain act constituted "negligence" or otherwise fell on the wrong side of some line, then this is the result of using a particular language and of looking at some things rather than others. Together, this turns the act into rule breaking behavior and the involved practitioner into a rule breaker.

A few years ago, my wife and I went for dinner in a neighboring city. We parked the car along the street, amid a line of other cars. On the other side of the street, I saw a ticket machine, so I duly went over, put some cash in the machine, got my ticket, and displayed it in the car windshield. When we returned from dinner, we were aghast to find a parking ticket the size of a half manila envelope proudly protruding from under one of the wipers. I yanked it away and ripped it open. Together we poured over the fine print to figure out what on earth we had violated. Wasn't there a ticket in our car windshield? It had not expired yet, so what was going on? It took another day of decoding arcane ciphers buried in the fine to find the one pointing to the exact category of violation. It turned out that it had somehow ceased to be legal to park on that side of that particular stretch of that street sometime during our dinner on that particular evening. I called a friend who lives in this city to get some type of explanation (the parking police answered the phone only with a taped recording, of course).

My friend must have shaken his head in blend of disgust and amusement. "Oh, they do this all the time in this town," he acknowledged. "If it hasn't been vandalized yet, you may find a sign the size of a pillow case suspended somewhere in the neighborhood, announcing that parking on the left or right side of the street is not permitted from like six o-clock until midnight on the third Tuesday of every month except the second month of the fifth year after the rule went into effect. Or something."

I felt genuinely defeated (and yes, we paid our fine). A few weeks later, I was in this city again (no, I did not park; I no longer dared to), and indeed found one of the infamous exception statements, black letters on a yellow background, hovering over the parking bays in a sidewalk. "No parking 14–17 every third of the second month," or some such vague decree.

This city, I decided, was a profile in the construction of offense. Parking somewhere was perfectly legal one moment and absolutely illegal the next. The very same behavior, which had appeared so entirely legitimate at the beginning of the evening (there was a whole line of cars on that side of the street, after all, and I did buy my ticket), had morphed into a violation, a transgression, an offense inside the space of a dinner. The legitimacy, or culpability of an act, then, is not inherent in the act. It merely depends on where we draw the line. In this city, on one day (or one minute), the line is here. The next day or minute, it is there. Such capriciousness must be highly profitable, evidently. We were not the only car left on the wrong side of the road when the rules changed that evening. The whole line that had made our selection of a parking spot so legitimate was still there—all of them bedecked with happily fluttering tickets. The only ones who could really decrypt the pillow-case size signs, I thought, were the ones who created them. And they probably planned their ambushes in close synchronicity with whatever the signs declared.

I did not argue with the city. I allowed them to declare victory. They had made the rules and had evolved a finely tuned game of phasing them in and out as per their intentions announced on those abstruse traffic signs. They had spent more resources in figuring out how to make money off of this than I was willing to invest in learning how to beat them at their own game and avoid being fined (I will take public transport next time). Their construction of an offense got to reign supreme. Not because my parking suddenly had become "offensive" to any citizen of that city during that evening (the square and surrounding streets were darker and emptier than ever before), but because the city decided that it should be so. An offense does not exist by itself "out there," as some objective reality. We (or prosecutors, or city officials) are the ones who construct the offense—the willful violation, the negligence, the recklessness.

What you see as rule breaking action, then, and how much retribution you believe it deserves, is hardly a function of the action. It is a function of your interpretation of that action. And that can differ from day to day or minute to minute. It can change over time, and differ per culture, per country. It is easy to show that the goal posts for what counts as a rule breaking shift with time, with culture, and also with the outcome of the action. The "crimes" I deal with in this book are a good example. They are acts in the course of doing normal work, "committed" by professionals—nurses, doctors, pilots, air traffic controllers, policemen, managers. These people have no motive to kill or maim or otherwise hurt, even though their job gives them both the means and opportunities to do so. Somehow, we manage to convert normal acts that (potentially) have bad outcomes into rule breaking or even criminal behavior. The kind of behavior that we believe should be punished. This, though, says labeling theory, hinges not on the essence of the action. It hinges on our interpretation of it, on what we call the action.

CONTROL THEORY

There are several ways that control theory can help you understand why people break rules. Let's first look at people's control of their own actions. Control, or rational choice theory, suggests that people make a decision to break a rule based on a systematic and conscious weighing of the advantages and disadvantages of doing so. An advantage might be getting to the goal faster, or being able to take a shortcut. One important disadvantage is getting caught and facing the consequences. If the probability of getting caught is low, the theory predicts that the likelihood of breaking the rule goes up. In other words, if people get away with it, they will break the rule.

Control theory suggests that stronger *external* control of behavior is a way to make people follow the rules. As long as such external control is lax, or nonexistent, it will be no surprise if people don't comply. One thing that we often need for control theory to work is the opportunity to break the rules. This is another important part of theorizing: people will not break the rules unless given the opportunity to do so.

Cameras have expanded into an area where a mistake can mean the difference between life and death: operating rooms. A medical center has installed cameras in all 24 of its operating rooms (ORs), which performed nearly 20,000 surgeries in 2013. With the national average of approximately 40 wrong-site surgeries and about a dozen retained surgical objects left in patients every week, the new pilot program at the hospital strengthens patient safety by providing hospitals with real-time feedback in their ORs. They call it remote video auditing (RVA) in a surgical setting.

RVA ensures that surgical teams take a "timeout" before they begin a procedure. The team then goes through a patient safety checklist aimed at avoiding mistakes. Each OR is monitored remotely once every two minutes to determine the live status of the procedure and ensure that surgical teams identify and evaluate key safety measures designed to prevent "never events," such as wrong-site surgeries and medical items inadvertently left in patients. The cameras also are used to alert hospital cleaning crews when a surgery is nearing completion, which helps to reduce the time it takes to prepare the OR for the next case. To reduce the risk of infections, the monitoring system also confirms whether ORs have been cleaned thoroughly and properly overnight. In addition, all staff can see real-time OR status updates and performance feedback metrics on plasma screens throughout the OR and on smart-phone devices. In a matter of weeks, patient safety measures improved to nearly perfect scores.

The executive director of the hospital reported that within weeks of the cameras' introduction into the ORs, the patient safety measures, sign-ins, time-outs, sign-outs, as well as terminal cleanings all improved to nearly 100 percent, and that a culture of safety and trust was now palpable among the surgical team.[47]

If you believe that better monitoring and surveillance give you control over what people do, and determine whether they break rules or not, then that says a number of things about both you and the people you believe need such control.

- People are not internally motivated to follow the rules. They need prodding and checking to do so.
- You believe, as the East Germans liked to say, that "trust is good; control is better."
- The reason your people don't follow the rules is that they have made a rational calculation that the advantages of breaking the rules outweigh the disadvantages. You can change that calculus by increasing the probability of getting caught.
- Having someone (or more likely, something) watch over them will get them to follow the rules, whether that someone (or something) is actually looking at them specifically or not.

Research has established that people who are used to being watched change their behavior as if they are monitored, even when they are not. This was the principle behind the so-called panopticum, the circular prison designed by Jeremy Bentham in the late eighteenth century. In this prison, inmates were placed in cells that circled a central observation post. From that post, the staff could watch all the inmates (hence the name panopticum, or *see all*). Bentham thought that this basic plan was equally applicable to hospitals, schools, sanatoriums, daycare centers, and asylums. It has, however, mostly influenced prisons around the world.

Cameras (like those used in RVA in the preceding example), recording devices (cockpit voice recorders, for example), or other monitoring systems (intelligent vehicle monitoring systems, for example) all represent panoptica as well. The idea is that they can (potentially) observe everything, anytime. Of course, they don't. Recall from the preceding example, for instance, that the OR gets sampled by the RVA every two minutes. What if the "time-out" procedure takes 119 seconds and falls between the two samples? Those who believe in a panopticum will say that it doesn't matter: as long as people believe that they are being watched, or that they *can* be watched, they will change their behavior.

But do people change their behavior in the way that you want?

- By putting in a camera, you cannot make people more committed to doing the work the way you want them to. You might just make them more afraid of the consequences if they don't.
- You also make very clear that you do not trust your people, and that they can't trust you, except that you will take action if they exceed the norms you find important.
- Perhaps that is good enough for you. But people are good at finding ways to make it look as if they are doing what you want them to.

With this sort of intervention, your people might not change their behavior or motives as much as changing the way it looks to the outside. And it certainly does contribute to a mature relationship of trust between you and the people you are watching.

LEARNING THEORY

People do not come to work to break rules; they typically come to work to get things done. But in doing this, they need to meet different goals—at the same time. This creates goal conflicts. For example, they need to achieve a particular production target or a deadline. Or save resources. But they also need to do so safely. Or within a budget. And within regulations and rules that apply. In many cases, it is hard to do all at the same time. In the pursuit of multiple incompatible goals, and under the pressure of limited resources (time, money), something will have to give. That something can be the rules that apply to the job. Gradually people come to see rule breaking behavior or system performance as normal or acceptable. Or expected even. Here is a great example of this.

The Columbia *space shuttle accident focused attention on the maintenance work that was done on the shuttle's external fuel tank. Maintenance workers were expected to be safe, follow the rules for reporting defects, and get the job done. A mechanic working for the contractor, whose task it was to apply the insulating foam to the external fuel tank, testified that it took just a couple of weeks to learn how to get the job done, thereby pleasing upper management and meeting production schedules. An older worker soon showed him how he could mix the base chemicals of the foam in a cup and brush it over scratches and gouges in the insulation, without reporting the repair.*

The mechanic found himself doing this hundreds of times, each time without filling out the required paperwork. This way, the maintenance work did not hold up the production schedule for the external fuel tanks. Inspectors often did not check. A company program that once had paid workers hundreds of dollars for finding defects had been watered down, virtually inverted by incentives for getting the job done now.

Learning theory suggests that people break a rule because they have learned that there are no negative consequences, and that there are, in fact, positive consequences. Some have called this fine tuning—fine tuning until something might break:

> Experience with a technology may enable its users to make' fewer mistakes and to employ the technology more safely, and experience may lead to changes in hardware, personnel, or procedures that raise the probability of success. Studies of industrial learning curves show that people do perform better with experience. Better, however, may mean either more safely or less so, depending on the goals and values that guide efforts to learn. If better means more cheaply, or quicker, or closer to schedule, then experience may not raise the probability of safe operations.[48]

Success makes subsequent success appear more probable, and failure less likely. This stabilizes people's behavior around what you might call rule breaking. But consider it from their angle. If a mechanic in the preceding example would suddenly start filling out all the required paperwork, what would happen? Colleagues would look strangely at her or him, production would be held up, and managers might get upset. And the mechanic might quickly lose her or his job. Fine tuning, or such a normalization of deviance, turns rule breaking into the new norm. If you don't do it, then *you* are actually the deviant. And your "deviance" (or "malicious compliance" with the rules) is seen as noncompliant, or nonconformant with the rules (the informal, unwritten, peer group rules) by which the system runs. Nuclear power plant operators have in fact been cited for "malicious compliance." They stuck to the rules religiously, and everything in the operation pretty much came to a standstill.

Nick McDonald, a leading human factors researcher at Trinity College Dublin, and his colleagues investigated the "normal functioning" of civil aircraft maintenance and found much of this confirmed. "Violations of the formal procedures of work are admitted to occur in a large proportion (one third) of maintenance tasks. While it is possible to show that violations of procedures are involved in many safety events, many violations of procedures are not, and indeed some violations (strictly interpreted) appear to represent more effective ways of working.

Illegal, unofficial documentation is possessed and used by virtually all operational staff. Official documentation is not made available in a way which facilitates and optimizes use by operational staff.

The planning and organizing of operations lack the flexibility to address the fluctuating pressures and requirements of production. Although initiatives to address the problems of coordination of production are common, their success is often only partial.

A wide range of human factors problems is common in operational situations. However, quality systems, whose job it is to assure that work is carried out to the required standard, and to ensure any deficiencies are corrected, fail to carry out these functions in relation to nontechnical aspects of operations. Thus, operational performance is not directly monitored in any systematic way; and feedback systems, for identifying problems which require correction, manifestly fail to demonstrate the achievement of successful resolution to these problems.

Feedback systems to manufacturers do not deal systematically with human factor issues. Formal mechanisms for addressing human needs of users in design are not well developed.

Civil aviation, including the maintenance side of the business, is a safe and reliable industry. To this extent, the system is achieving its goals. How is this compatible with the statement given above? What we have to explain is a little paradoxical: on the one hand the system appears to malfunction in several important respects, yet this malfunctioning, being often a normal and routine fact of life in maintenance organizations, does not in itself imply that the system is not safe. On the other hand, it does indicate important vulnerabilities of the system which have been demonstrated in several well investigated incidents. Can the same set of principles explain at the same time how safety is normally preserved within a system and how it is compromised?"[49]

Learning theory also covers the other kind of learning identified in McDonald's research. This is the kind that is the result of interaction with others, possibly veterans of the practice or profession. They engage in informal teaching (like in the space shuttle external tank chemicals-in-a-cup example, but also as communicated in illegal documentation commonly used in maintenance, often called black books). Informal teaching, or the teaching of a "hidden curriculum," touches on those things that a profession would rather not talk about too openly, or that managers don't want to hear about. It might have a "teacher" open a task with something like: *Here, let me show you how it's done.* That not only hints at the difficulty of getting the job done within the formal rules, but also at goal conflicts, at the need to do multiple things at the same time. Informal teaching offers a way of managing that goal conflict, of getting around it, and getting the job done. As McDonald has shown, such skills are often hugely valued, with many in the organization not even knowing about them. A hidden curriculum shows that the world in which your people work is too complex to be adequately covered by rules. There are too many subtleties, nuances, variations, and smaller and larger difficulties. And there are too many other expectations. Following the rules is just yet another thing your people have to try to accomplish.

THE BAD APPLE THEORY

But can this help breed evil or bad practitioners? Many people believe that a just culture should pursue and sanction practitioners who come to work to do a good job yet make the occasional inadvertent error. But what about the practitioners who consistently get complaints or get things wrong? Don't you then have a responsibility to deal with these "bad apples" decisively and effectively? Some seem to have experiences that tell them so. Let's revisit the questions raised in the Preface—about the successes and limits of the systems approach. It was very important to convince policymakers, hospital chiefs, and even patients that the problem was not bad doctors, but systems that needed to be made safer:

> Individual blame was deemed the wrong solution to the problem of patient safety; as long as specific individuals were deemed culpable, the significance of other hazards would go unnoticed. The systems approach sought to make better diagnosis and treatment of where the real causes of patient safety problems lay: in the "latent conditions" of healthcare organisations that predisposed to error. In order to promote the learning and commitment needed to secure safety, a "no-blame" culture was advocated. With the spotlight switched off individuals, the thinking went that healthcare systems could draw on human factors and other approaches to improve safety.
>
> Almost certainly, the focus on systems has been an important countervailing force in correcting the long-standing tendency to mistake design flaws for individual pathologies. There can be no doubting the ongoing need to tackle the multiple deficits in how healthcare systems are designed and organised. Encouraging examples of just how much safety and other aspects of quality can be improved by addressing these problems continue to appear. Yet, recent years have seen increasing disquiet at how the importance of individual conduct, performance and responsibility was written out of the patient safety story... We need to take seriously the performance and behaviours of individual clinicians if we are to make healthcare safer for patients.
>
> One study found that a small number of doctors account for a very large number of complaints from patients: 3% of doctors generated 49% of complaints, and 1% of doctors accounted for 25% of all complaints. Moreover, clinician characteristics and past complaints predicted future complaints. These findings are consistent with other recent research, including work showing that some doctors are repeat offenders in surgical never-events, and a broader literature that has explored the phenomenon of "disruptive physicians" with behaviour problems as well as those facing health or other challenges that impact on patient care.
>
> These studies show that a very small number of doctors may contribute repeatedly not just to patient dissatisfaction, but also to harm and to difficult working environments for other healthcare professionals. Identifying and dealing with doctors likely to incur multiple complaints may confer greater benefit than any general strategy directed at clinicians in general.[7]

There is no doubt about the genuine concern for patient safety expressed in these ideas. People's experiences can indeed lead them to want to get rid of certain hospital staff. One can understand the seduction of sanctioning noncompliant doctors or getting rid of the deficient practitioners—the system's bad apples—altogether.

In 1925, German and British psychologists were convinced they had cracked the safety problem in exactly this way. Their statistical analysis of five decades had led them to accident-prone workers, misfits whose personal characteristics predisposed them to making errors and having accidents. Their data told the same stories flagged by Levitt: if only a small percentage of people are responsible for a large percentage of accidents, then removing those bad apples will make the system drastically safer.

It didn't work. The reason was a major statistical flaw in the argument. For the accident-prone thesis (or bad apple theory) to work, the probability of error and accident must be equal across every worker or doctor. Of course it isn't. Because they engage with vastly different problems and patient groups, not all doctors are equally likely to harm or kill patients, or get complaints. Personal characteristics do not carry as much explanatory load for why things go wrong as context does. Getting rid of Levitt's 3% bad doctors (as measured by complaints and adverse events) may simply get rid of a group of doctors who are willing to tackle trickier, more difficult cases. The accident-prone thesis lived until World War II, when the complexity of systems we made people work with—together with its fatal statistical flaw—did it in. As concluded in 1951:

> ...the evidence so far available does not enable one to make categorical statements in regard to accident-proneness, either one way or the other, and as long as we choose to deceive ourselves that they do, just so long will we stagnate in our abysmal ignorance of the real factors involved in the personal liability to accidents.[50]

Is there any point in reinvoking a failed approach to safety that was debunked in 1951? Errors are not the flaws of morally, technically, or mentally deficient "bad apples," but the often predictable actions and omissions that are systematically connected to features of people's tools and tasks.[29]

Bad apples might still be a concern. Even the European Union has recently created a "black list"[51] of putatively deficient medical practitioners. Of course some practitioners should not be allowed to treat patients or control airplanes.

- But who let them in?
- Who recruited them, trained them?
- Who mentored them, promoted them, employed them, supervised them?
- Who gave them students to work with, residents to educate?
- Who allowed them to stay?
- What are they doing that is actually good for other organizational goals (getting the job done where others wouldn't, even if it means playing fast and loose with the rules)?

If we first start to worry about incompetent practice once such practitioners are comfortably ensconced and have been doing things for years or decades, we are way behind the curve. The question is not how we get rid of bad apples, but in what way we are responsible for creating them in the first place. Yet that is not the question asked by those concerned with "bad apples" in their midst:

> Many countries, including the USA and the UK, have introduced periodic recertification or "revalidation" of doctors in an attempt to take a systematic, preventative, risk-based approach. In theory, relicencing should pick up doctors whose practice is unsafe, and ensure they are enabled to improve or that their licences are restricted or removed. But relicencing systems are remarkably difficult to design and operate, not least because it is hard to ensure that bad apples are detected (and appropriate action taken) while also encouraging good apples to thrive. Any regulatory system of "prior approval" involves a high regulatory overhead, typically imposing high burdens both on the regulatees (individuals and organisations) as well as on the regulators. The "good apples" may have to divert their time and resources in demonstrating that they are good, while the bad apples may find ways of evading detection. Therefore, it is perhaps unsurprising that relicencing regimes for doctors typically attract a high level of regulatee complaint, both about burden and lack of efficacy.[7]

For sure, there are individual differences in how competent or "fit" people are for certain kinds of work. And in some domains, structures to oversee and regulate competence may currently not be as effective as in some other worlds. But is it the "bad apple's" responsibility for being mismatched with his or her work? Or is that a system responsibility, a managerial responsibility? Perhaps these are systems that need to be improved, and such oversight is a system responsibility. Other worlds have had less trouble asserting such oversight—from the beginning of a career to the end—and having no trouble getting practitioners in the domain to accept it (see Table 2.1).

What appears on the surface as a simple competence problem often hides a much deeper world of organizational constraints, peer and patient expectations and assumptions, as well as cultural and professional dispositions. A simple comparison between two different safety-critical fields, aviation and medicine, shows the most outward signs of vastly different assumptions about competence and its continuity, assurance, and maintenance.

For example, training on a new piece of technology, and checking somebody's competency before she or he is allowed to use it in practice, is required in aviation. It makes no assumptions about the automatic transference of basic flying or operating skills. Similarly, skill maintenance is ensured and checked in twice-yearly simulator sessions, necessary to retain the license to fly the airplane. Failing to perform in these sessions is possible and entails additional training and rechecks.

TABLE 2.1
Different Assumptions about Competency and How to Ensure and Maintain It over Time in Aviation and Medicine

Ideas about Competence	Aviation	Medicine
Type training and check before use of new technology	Always	No
Recurrent training in simulator for skill maintenance	Multiple times a year	Still haphazard
Competency checks in simulator	Twice a year	
Emergency training (both equipment and procedures)	Every year	Not often
Direct observation and checking of actual practice	Every year or even more regular	Not so often, not regular
Crew resource management training	Every year	Not standard
Standard-format briefing before any operational phase	Always	Hardly
Standardized communication/phraseology	Yes	No
Standardized procedures for accomplishing tasks	Yes	No
Standardized divisions of labor across team members	Yes	No
Extensive use of checklists	Yes	No
Duty time limitations and fatigue management	Yes	Grudgingly
Formalized and regulated risk assessment before novel operation	Yes	Hardly ever

Note: The table is a generalization. Different specialties in medicine make different assumptions and investments in competence, and there is a slow but gradual move toward more proficiency checking, checklist use, and teamwork and communication training in medicine too.[52]

Other aspects of competence, such as the ability to collaborate on a complex task in a team, are tackled by the use of standard communication and phraseology, standard-format briefings before each new operational phase, standard procedures and divisions of labor for accomplishing operational tasks, and the extensive use of checklists to ensure that the work has been done. Finally, there are limits on duty time that take into account the inevitable erosion of skills and competence under conditions of fatigue. Novel operations (e.g., a new aircraft fleet or a new destination) are almost always preceded by a risk assessment before their approval and launch, by airline and regulator alike. Skills or competence or safety levels demonstrated in one situation are not simply believed to be automatically transferable to another.

Competence alone is not trusted to sustain itself or to be sufficient for satisfactory execution of safety-critical tasks. Competence needs help. Competence is not seen as an individual virtue that people either possess or do not possess on entry into the profession. It is seen as a systems issue, something for which the organization, the regulator, and the individual all bear responsibility for the entire lifetime of the operator.[52] Indeed, it would seem that getting rid of the "bad apples," or putatively incompetent practitioners at the back end, is treating a symptom. It does not begin to delve into, let alone fix, a complex set of deep causes. This, of course, is one of the main issues with a retributive just culture as a response to rule violations. It targets the symptom, not the problem. And by targeting the symptom (i.e., the rule violator), it sends a clear message to others in the organization: *Don't get caught violating our rules; otherwise you will be in trouble too.* But work still needs to get done.

STUPID RULES AND SUBCULTURE THEORY

Some rules are actually stupid. That is, they do not fit the work well or at all. Or they get in the way of doing the work. Stupid rules cost time and money and don't help control risk. This makes it unsurprising that responsible practitioners will find ways to get their work done without complying with such rules.

An aviation business had strong internal controls to manage access to its expensive spare parts inventory. This seemed to make sense. Just like hospital pharmacies have such access controls. The controls appeared reasonable, because they were protecting a valuable asset. But it did mean that maintenance and engineering staff were walking more than they were working. It would take them up to four hours of walking each day to cover the almost ten miles to collect the paperwork, the signatures, and finally the parts vital to their maintenance work. Similarly, a large construction company implemented significant new controls, systems, and processes to reduce the risk of project overrun. In isolation, all the steps looked reasonable. But together, they created more than 200 new steps and handoffs in the approval process. The result was that it took up to 270 days to obtain "approval for approval." Information would get dated, and the process would often have to be revisited if it was to stay at all accurate. The overruns that the organization had tried to reign in had been considerably cheaper than the productivity losses and bureaucracy costs created by the new rules, systems, and processes. The medicine was literally worse than the disease.

Then there is the company that makes its people await approval for small taxi fares from a weekly executive team meeting; the firm that rejects application forms from potential customers if they are not filled in using black ink; the business that demands from its receptionists to record every cup of coffee poured for a visitor, yet allows managers to order as much alcohol as they like; and the organization that insists staff complete an ergonomic checklist and declaration when they move desks, and then introduces hot-desking so that everyone spends 20 minutes a day filling out forms. So yes, there actually are stupid rules.[53]

Groups of people might even develop subcultures in your organization that find ways of getting work done without letting such rules get in their way. From the outside, you might see this as a subculture that considers itself above those rules. Sometimes you might even see this as a "can-do culture." Can-do cultures are often associated with people's ability to manage large technical or otherwise complex systems (e.g., NASA space shuttles, financial operations, military missions, commercial aircraft maintenance) despite obvious risks and vulnerabilities. What appears to set a can-do culture apart is the professional pride that people inside the subculture derive from being able to manage this complexity despite the lack of organizational understanding, trust, and real support. "Can-do" is shorthand for "Give us a challenge and don't give us the necessary resources; give us rules that are contradictory, silly, or unnecessary, and we can still accomplish it." The challenge such a subculture makes its own is to exploit the limits of production, budgets, manpower, and capacity. The challenge is to outwit the bureaucracy, to have it by the nose without it noticing.

Goal conflicts are an important factor here. Over the years, people in a subculture become able to prove to themselves and their organization that they can manage such contradictions and pressures despite resource constraints and technical and procedural shortcomings. They also start to derive considerable professional pride from their ability to do so. Subculture theory says that people on the inside identify more strongly with each other than with those in the organization above, below, or around them. They might talk disparagingly of their own managers, or of "B" team members, as people who don't get it, who don't understand, who can't keep up. Subcultures are also powerful in the kind of coaching and informal teaching that shows newer workers how to get work done.

Having worked with a number of air traffic control organizations (or air traffic navigation service providers, as they are formally known), I have learned that there are commonly "A" and "B" teams (or at least that) both inside and across different control towers and centers. Everybody basically knows who is part of which team, and such teams develop strong subcultural identities. These identities express themselves in the style in which people work and collaborate. An "A" team will typically have what is known as a "can-do culture." One evening, in one such team, the supervisor handed a controller a list of flights. A number of lines were highlighted with a marker pen. On being asked why these flights were so special, the controller explained that the accepted capacity for the coming hour was greater than the "official" allowable number of flights. She also explained that this was typical. The highlighted flights were the ones departing from the main airport that evening: if anything went wrong, the controller explained, those were the ones they would keep on the ground first. That way, any disruptions in a system that was being asked to run over its stated capacity could be dealt with "safely." For the controller and the supervisor, this was not only entirely nonproblematic. They were proud to be able to locally solve, or deal with, such a capacity challenge. They were proud of being able to create safety through their practice, despite the pressures, despite the lack of resources. Paradoxically, there may be some resistance to the introduction of tools and technical support (and even controller capacity) that would make people's jobs easier—tools or extra people that would provide them with more resources to actually meet the challenges posed to them. This resistance is born in part of professional pride: put in the tools, and the main source of professional pride would be threatened.

It might be seductive to tackle the problem of subcultures by trying to disband or repress that entire culture. But you might be attacking the goose that lays your golden eggs. "A" teams, or can-do subcultures, have become able, better than others, to generate more production, more capacity, more results. And they do so with the same or fewer resources than others, and probably with fewer complaints. What is interesting is that people in the subculture start to make the organization's problems (or your problems) their own problems. *They* want to solve the production challenge, *they* want to meet the targets, *they* want to overcome your organization's capacity constraints. And if they don't, subculture members might feel that they have disappointed not just their organization or its management, but also themselves and their immediate peers. They feel this even though these are not their personal targets to meet or problems to overcome. But they feel such a strong identification with that problem, and derive so much personal pride from solving it, within the norms and expectations of their subculture, that they will behave as if it is indeed their own problem.

When a subculture plays fast and loose with your rules it's because it has to. *You*, after all, expect people to achieve goals other than rule compliance too—such as producing, being efficient, fast, and cheap. And the nature of your operation allows them to compare themselves with others, by production numbers or other measurable targets, for instance.

In one nuclear power plant I visited, for example, the daily kilowatts produced by each of its three reactors were shown in bright lights over the entrance to the site. Reactors tend to develop individual "personalities" and thus allow the teams managing them to develop distinct cultures in how they get the most out of their reactor. This despite the fact that the nuclear industry has, since the Three Mile Island incident in 1979, embraced the notion (and cultivated the public image) that it is safe and productive precisely (if not only) because its people are compliant with the rules.

A smarter way to address a subculture problem (if that is what you think you have) is to look at the systems, rules, and processes that have created the conditions for subcultures to emerge in the first place. Take an honest, close look at the front end where the rules enter the system, where people have to do your organization's work while also complying with all those rules. Take a close look at the rules that you are asking your people to follow while expecting them to live up to your explicit and implicit production or other targets. For example, look for[53]

- **Overlap in the rules**. Rules often look at risks in isolation. There might be rules for safety, but also rules for security, for example. Or rules for process safety and personal safety. In certain cases these can conflict and create a compliance disaster. Subcultural responses to this are those that sort of show compliance with both, or at least make it look that way—based on what people themselves have learned about the nature of the risks they face in their work.
- **Too many cooks**. Rules that people need to comply with come from many places and directions, both internal and external to your organization. If these are not coordinated, overlap and contradiction, as well as sheer volume of rules, can all go up.
- **Lack of independence**. Rule-making and rule enforcement in most organizations are governed by the same unit. This gives the rule enforcers a serious stake in their compliance, because they themselves made the rule.
- **Micro-management**. Rules are often unnecessarily detailed or overspecified. The goal might be to meet all the possible conditions and permutations in which they apply. But of course this is impossible. The world of work is almost always more complex than the rules we can throw at it.
- **Irrelevance**. Rules can become obsolete as work progresses or new technologies become available. Decision makers should not "fire and forget" the rules into an organization. Consider putting a "sunset" or expiration data on the rules you create: don't leave them as rigid markers of long-gone management decisions.
- **Rule creep**. Driven in part by those who enforce them and those who believe in how they control risk, rules tend to creep. They seep into areas and activities that were not their original target, asking people to comply with things that may leave them puzzled.
- **Excessive reporting or recording requirements**. Some processes, systems, and rules demand excessive information. This becomes especially disruptive when the compliance and reporting demands are not well-coordinated, leading to overlap or duplication.
- **Poor consultation**. Many rules get made and enforced by those who don't do the actual work themselves, or no longer do. This is what consultation with those who do the work is supposed to solve. But consultation often gets done poorly or merely (and ironically) to "comply" with consultation rules and requirements.
- **Overreactions**. High-visibility problems or incidents tend to drive managers to decisions: something needs to be done. Or they need to be seen to do something. New rules and procedures are a common way to do so. But it doesn't mean they help on the ground, at the coal face, at all.
- **Slowness**. The granting of approvals (e.g., getting a lock-out/tag-out for doing work on a process plant) can take too long relative to immediate production goals. This can encourage the formation of can-do subcultures that get your work done in spite of you and your rules.

I was visiting a remote construction site not long ago. All workers, spread out over a flat, hot field, were wearing their hard hats. These were the rules, even though there was nothing that could fall from above, other than the sky itself. Any signs of noncompliance might have to be sought in more subtle places. I noticed how workers' hard hats had all kinds of stickers and decals and codes on them. Some indicated the authorization to use mobile phones in certain restricted areas; others showed that the wearer was trained in first aid. Then I spotted a supervisor who had "GSD" pasted on the back of his hard hat. I asked what it stood for, expecting some arcane permission or authority.

"Oh, that means he 'Gets Stuff Done,'" I was told. And "stuff" was actually not the word used. Intrigued, I spent some more time with the team. This was a can-do man. A supervisor leading a subcultural "A" team, which G..S..D… They certainly complied with the organization's often unspoken wish to G..S..D…, if not with everything else.

RESILIENCE THEORY

This brings us to resilience theory. This theory says that work gets done—and safely and productively so—not because people follow rules or prescriptive procedures, but because they have learned how to make trade-offs and adapt their problem solving to the inevitable complexities and contradictions of their real world. They typically do this so smoothly, so effectively, that the underlying work doesn't even go noticed or appreciated by those in charge. Leaders may persist in their belief that work goes well and safely because their people have listened to them and are following the rules:

> When an organization succeeds, its managers usually attribute this success to themselves, or at least to their organization. The organization grow[s] more confident of their managers' skill, and of existing programmes and procedures.[48]

The smoothness and unintrusive way in which people in the organization adapt and work can offer managers the impression that safety is preserved because of stable application of rules. Even though—largely invisibly—the opposite may be true.

> In day-to-day operations, these adaptive capacities tend to escape attention as well as appreciation, also for the contribution to safety. As put by Weick and Sutcliffe (2001)[54]: "Safety is a dynamic non-event.… When nothing is happening, a lot is happening." Adaptive practices (resilience) are rarely part of an organization's official plan or deliberate self-presentation, maybe not even part of its (managerial) self-understanding. Rather it takes place "behind the rational facades of the impermanent organization."[55]

When nothing is happening, that is because *a lot is happening*. A lot more than people complying with rules in any case. In trying to understand the creation and nature of safety in organizations, resilience theory asks the question "why does it work?" rather than "why does it fail?" Things go right, resilience theory argues, not because of adequate normative prescription from above, and people complying with rules, but because of the capacity of your people to recognize problems, adapt to situations, interpret procedures so that they match the situation, and contribute to margins and system recovery in degraded operating conditions.[56]

The question of "why things go right" is strangely intriguing to safety people, as they are mostly interested in cases in which safety was lost, not present. But most work goes well, after all. Only a tiny proportion goes wrong. It would be ignorant to believe that rules are broken only in that small proportion, and that things otherwise go right because people strictly comply with all the rules. And you probably know this already, even if you didn't realize it. Just consider the following:

- Virtually all organizations **reward seniority**. The longer a person has practiced a profession (flying a plane, operating on patients, running a reactor), the more he or she is rewarded—in status, economically, rostering, and such. This is not only because he or she has simply stuck it out for longer. Organizations reward seniority because with seniority comes experience. And with experience comes judgment, insight, and expertise—the kind of discretion that allows people to safely make trade-offs and adaptations so as to meet multiple organizational and task goals simultaneously, even if these conflict.
- The **work-to-rule strike** (one option for industrial action) shows that if people religiously follow all the rules, the system comes to a grinding halt. In air traffic control, this is easy to demonstrate, for instance. Nuclear power has a different phrase for this: *malicious compliance*. But how can compliance be malicious, if compliance is exactly what the organization demands? And how can following all the rules bring work to halt? It can, says resilience theory, because real work is done at the dynamic and negotiable interface between rules that need following and complex problems and nuanced situations that need solving and managing. In this negotiation, something needs to give. Rules get situated, localized, interpreted. And work gets done.

When it asks, "why does it work?" resilience theory sees people not as a problem to control, but as a resource to harness. Things work not because people are constrained, limited, controlled, and monitored, but because they are given space to do their work with a measure of expertise, discretion—innovation even. There is an interesting link between the trust that should be part of your just culture and your organization's understanding of how work actually gets done:

> Firms with a lack of trust in their employees limit their use of judgment or discretion. Large organizations, in particular, tend to frown on discretion, meaning that many employees—when faced with a divergent choice between doing the right thing and doing what the rule says—will opt for the latter. They prefer the quiet life of "going by the book," even if they doubt the book's wisdom.[53]

Cultures of compliance seem to have become more popular than cultures of trust, learning, and accountability. A culture of compliance is a culture that puts its faith in fixed rules, bureaucratic governance, paper trails, and adherence to protocol and hierarchy rather than local expertise, innovation, and common sense. A 30-year veteran signal engineer at a railway company told me not long ago how he is about to resign from his job. He could no longer put up with the bureaucratic "crap" that made him fill out a full 40 minutes worth of paperwork before he could get onto the tracks to fix a signaling issue. The irony riled him: he was complying with checklists and paper-based liability management in the name of safety, while hundreds of innocent passengers were kept at risk during that very time, because of an unfixed signal! He scoffed at the paperwork, its duplication and stupidity, as it was all written by those who'd never been on the track themselves. Would it be "just" to sanction this signal engineer—who has been on and off the tracks for three decades—to enter a track without all the paperwork done so that at least he might assure the safety of the company's customers?

The signal engineer is not alone. Similar cultures of compliance have risen almost everywhere, with consequences for productivity and safety. In Australia, for example, the time required for employees to comply with self-imposed (as opposed to government-imposed) rules has become a big burden: staff members such as the signal engineer spend an average 6.4 hours on compliance each week, and middle managers and executives spend an average of 8.9 hours per week. Whereas the compliance bill for government-imposed rules came to $94 billion in 2014, the cost of administering and complying with self-imposed rules and regulations internal to an industry or organization came to $155 billion. As a proportion of the gross domestic product, each Australian working man and woman is spending eight weeks each year just covering the costs of compliance. Some do benefit, however. The proportion of compliance workers (those whose job it is to write, administer, record, and check on compliance with rules and regulations) has grown from 5.6% of the total workforce in 1996 to 9.2% in 2014. One in every eleven employed Australians now works in the compliance sector.[53]

If you don't trust your employees, then how can you ever build a just culture? Remember, a just culture is a culture of trust, learning, and accountability. Organizations such as those described in the preceding text might have only one of those three: accountability. They hold their people accountable for following the rules, for going by the book (or making it convincingly look as if they do). But they don't really trust their people, and they are not learning much of value from them. After all, the problems are all solved already. That is what the book is for. Apply the book; solve the problem in the way someone else has already figured out for you. But what if there is a creeping, yet persistent drift in how the operation starts to mismatch the rules that were once designed for it? This has happened in a number of cases that became spectacular accidents[57] and is probably happening in many places right now. In a culture without trust, without the sort of empowerment that recognizes your people as resources to harness, your people might not come forward with better or safer ideas for how to work. You don't learn, and you might just end up paying for it:

> Top executives can usually expect both respect and obedience from their employees and managers. After all, the executives have the power to fire them. But whether or not the leaders earn their trust is a different issue, and executives ignore the difference at

> their peril. Without trust, the corporate community is reduced to a group of resentful wage slaves and defensive, if not ambitious, managers. People will do their jobs, but they will not offer their ideas, or their enthusiasm, or their souls. Without trust the corporation becomes not a community but a brutish state...[58]

Fortunately, many leaders intuitively recognize that top-down, reductionist safety management through rule compliance is not enough. And that it may get in the way of doing work safely or doing work at all. They understand that there is something more that they need to recognize and nurture in their people that goes beyond their acknowledged or handed-down safety capabilities. They understand that sometimes *imagination* is a more important human faculty than *reason*. They understand that trust is perhaps less something they *have*, and more something they *do*. They understand that trust is an option, a choice for them to pursue or make. It is something they themselves help create, build, maintain—or undermine and break as part of the relationships they have with other people. They understand that it is not so much people who are trustworthy or not. It is, rather, their relationship with people that makes it so.[58] Here are some ways to begin rebuilding that trust, to invest in your people as a resource, rather than as a problem[53]:

- First get your people to **cleanse** their operations from stupid rules. Ask them what the absolute dumbest thing is that *you* asked them to do today in their work. If you really want to know and show no prejudice, they will tell you (indeed, they will give you their *account*). This shows your trust by allowing them judge what is dumb and what is useful. And you will learn some valuable things as well: things that may keep you from a productive and just culture.
- Then **change** the way you see people and hold them accountable. Hold them accountable not for compliance, because then you still see your people as a problem to control. Rather, hold them accountable for their performance, for their imagination, for their ideas. Hold them accountable by creating all kinds of ways for them to contribute their accounts, their stories, of how to make things work, of how to make things go right. That way, you see your people as essential resources to harness.
- Then **challenge** your people throughout the organization (including staff departments, frontline supervision, line management) to ask "what must go right?" instead of "what could go wrong?" That allows you to focus any remaining rules on what really matters. Make sure you ask whether you really need the rule or whether there is a better way to achieve the desired outcome. There probably is, but the challenge for you is to trust and allow your people to come up with it.

You would be in good company if you can pull this off. As Thomas Edison, inventor and founder of General Electric said, "Hell, there are no rules here... we're trying to accomplish something."[53]

CASE STUDY

Hindsight and Shooting Down an Airliner

There is a very important aspect to your judgment of your people's "rule breaking." And that is knowing the outcome of their actions. This knowledge has a huge influence on how you will (but of course shouldn't) judge the quality of their actions.

Years ago a student of mine was rushing to the scene of a terrible accident that happened to the airline he worked for right after takeoff on a dark, early morning. The crash killed dozens of people, including his pilot colleague. My student told my class that as he was sitting in his car, gripping the steering wheel in tense anticipation of what he was going to find, he repeated to himself: "Remember, there was no crash. An accident hasn't happened. These guys had just come to work and were going to spend a day flying. Remember..." He was bracing himself against the visual onslaught of the logical endpoint of the narrative—the debris of an airplane and human remains scattered through a field and surrounding trees. It would immediately trigger questions such as, How could this happen? What actions or omissions led up to this? What did my colleague miss? Who is to blame?

My student was priming himself to bracket out the outcome because otherwise all of his questions and assessments were going to be driven solely by the rubble strewn before his feet. He wanted to avoid becoming the prisoner of a narrative truth whose starting point was the story ending. The story ending was going to be the only thing he would factually see: a smoking hole in the ground. What preceded it, he would never see, he would never be able to know directly from his own experience. And he knew that when starting from the outcome, his own experiential blanks would be woven back into time all too easily, to form the coherent, causal fabric of a narrative. He would end up with a narrative truth, not a historical one, a narrative that would be biased by knowing its outcome before it was even told. His colleague didn't know the outcome either, so to be fair to him, my student wanted to understand his colleagues' actions and assessments not in the light of the outcome, but in the light of the normal day they had ahead of them.

We assume that if an outcome is good, then the process leading up to it must have been good too—that people did a good job. The inverse is true too: we often conclude that people may not have done a good job when the outcome is bad.

This is reflected, for example, in the compensation that patients get when they sue their doctors. The severity of their injury is the most powerful predictor of the amount that will be awarded. The more severe the injury, the greater is the compensation. As a result, physicians believe that liability correlates not with the quality of the care they provide, but with outcomes over which they have little control.[59]

Here is the common reflex: the worse the outcome, the more we feel there is to account for. This is strange: the process that led up to the outcome may not have been very different from when things would have turned out right. In fact, we often have to be reminded of the idea that an outcome does not, or should not, matter in how we judge somebody's performance. Consider the following quote: "If catnapping while administering anesthesia is negligent and wrongful, it is behavior that is negligent and wrongful whether harm results or not."[60] Also, quality processes can still lead to bad outcomes because of the complexity, uncertainty, and dynamic nature of work.

The Hindsight Bias

If we know that an outcome is bad, then this influences how we see the behavior that led up to it. We will be more likely to look for mistakes. Or even negligence. We will be less inclined to see the behavior as "forgivable." The worse the outcome, the more likely we are to see mistakes, and the more things we discover that people have to account for. Here is why.

- After an incident, and especially after an accident (with a dead patient, or wreckage on a runway), it is easy to see where people went wrong, what they should have done or avoided.
- With hindsight, it is easy to judge people for missing a piece of data that turned out to be critical.
- With hindsight, it is easy to see exactly the kind of harm that people should have foreseen and prevented. That harm, after all, has already occurred. This makes it easier for behavior to reach the standard of "negligence."

The reflex is counterproductive: like physicians, other professionals and entire organizations may invest in ways that enable them to account for a bad outcome (more bureaucracy, stricter bookkeeping, practicing defensive medicine). These investments may have little to do with actually providing a safe process.

Yet the hindsight bias is one of the most consistent and well-demonstrated biases in psychology. But incident reporting systems or legal proceedings—systems that somehow have to deal with accountability—have essentially no protections against it.

Lord Anthony Hidden, the chairman of the investigation into the devastating Clapham Junction railway accident in Britain, wrote, "There is almost no human action or decision that cannot be made to look flawed and less sensible in the misleading light of hindsight. It is essential that the critic should keep himself constantly aware of that fact."[61]

If we don't heed Anthony Hidden's warning, the hindsight bias can have a profound influence on how we judge past events. Hindsight makes us

- Oversimplify causality (this led to that) because we can start from the outcome and reason backwards to presumed or plausible "causes"
- Overestimate the likelihood of the outcome (and people's ability to foresee it), because we already have the outcome in our hands

- Overrate the role of rule or procedure "violations." Although there is always a gap between written guidance and actual practice (and this almost never leads to trouble), that gap takes on causal significance once we have a bad outcome to look at and reason back from
- Misjudge the prominence or relevance of data presented to people at the time
- Match outcome with the actions that went before it. If the outcome was bad, then the actions leading up to it must have been bad too—missed opportunities, bad assessments, wrong decisions, and misperceptions

If the outcome of a mistake is really bad, we are likely to see that mistake as more culpable than if the outcome had been less bad. If the outcome is bad, then there is more to account for. This can be strange, because the same mistake can be looked at in a completely different light (e.g., without knowledge of outcome) and then it does not look as bad or culpable at all. There is not much to account for. So hindsight and knowledge of outcome play a huge role in how we handle the aftermath of mistake. Let us look here at one such case: it looks normal, professional, plausible, and reasonable from one angle. And culpable from another. The hinge between the two is hindsight: knowing how things turned out.

The case study shows the difference between foresight and hindsight nicely—and how our judgment of people's actions depends on it.

Zvi Lanir, a researcher of decision making, tells a story of a 1973 encounter between Israeli fighter jets and a Libyan airliner. He tells the story from two angles: from that of the Israeli Air Force, and then from that of the Libyan airliner.[62] This works so well that I do that here too. Not knowing the real outcome, Israeli actions make good sense, and there would be little to account for. The decision-making process has few if any bugs.

- The incident occurred during daylight, unfolded during 15 minutes, and happened less than 300 kilometers away from Israeli headquarters.
- The people involved knew each other personally, were well acquainted with the terrain, and had a long history of working together through crises that demanded quick decisions.
- There was no evidence of discontinuities or gaps in communication or the chain of command.
- The Israeli Air Force commander happened to be in the Central Air Operations Center, getting a first-hand impression as events advanced.
- The Israeli chief of staff was on hand by telephone for the entire incident duration too.

The process that led up to the outcome, in other words, reveals few problems. It is even outstanding: the chief of staff was on hand to help with the difficult strategic implications; the Air Force commander was in the Operations Center where decisions were taken, and no gaps in communication or chain of command occurred. Had the outcome been as the Israelis may have suspected, then there would have been little or nothing to account for. Things went as planned, trained for, expected, and a good outcome resulted.

But then, once we find out the real outcome (the real nature of the Libyan plane), we suddenly may find reason to question all kinds of aspects of that very same process. Was it right or smart to have the chief of staff involved? What about the presence of the Air Force commander? Did the lack of discontinuities in communication and command chain actually contribute to nobody saying "wait a minute, what are we doing here?" The same process—once we learned about its real outcome—gets a different hue. A different accountability. As with the doctors that are sued: the worse the outcome, the more there is to account for. Forget the process that led up to it.

A Normal, Technical Professional Error

At the beginning of 1973, Israeli intelligence received reports on a possible suicide mission by Arab terrorists. The suggestion was that they would commandeer a civilian aircraft and try to penetrate over the Sinai Desert with it, to self-destruct on the Israeli nuclear installation at Dimona or other targets in Beer Sheva. On February 21, the scenario seemed to be set in motion. A sandstorm covered much of Egypt and the Sinai Desert that day.

At 13:54 hours (1:54 P.M.), Israeli radar picked up an aircraft flying at 20,000 feet in a northeasterly direction from the Suez bay. Its route seemed to match that used by Egyptian fighters for their intrusions into Israeli airspace, known to the Israelis as a "hostile route." None of the Egyptian war machinery on the ground below, supposedly on full alert and known to the Israelis as highly sensitive, came into action to do anything about the aircraft. It suggested collusion or active collaboration.

Two minutes later, the Israelis sent two F-4 Phantom fighter jets to identify the intruder and intercept it if necessary. After only a minute, they had found the jet. It turned out to be a Libyan airliner. The Israeli pilots radioed down that they could see the Libyan crew in the cockpit, and that they were certain that the Libyans could see and identify them (the Shield of King David being prominently displayed on all Israeli fighter jets).

At the time, Libya was known to the Israelis for abetting Arab terrorism, so the Phantoms were instructed to order the intruding airliner to descend and land on the nearby Refidim Airbase in the south of Israel. There are international rules for interception, meant to prevent confusion in tense moments where opportunities for communication may be minimal and opportunities for misunderstanding huge. The intercepting plane is supposed to signal by radio and wing-rocking, while the intercepting aircraft must respond with similar signals, call the air traffic control unit it is in contact with, and try to establish radio communication with the interceptor.

The Libyan airliner did none of that. It continued to fly straight ahead, toward the northeast, at the same altitude. One of the Israeli pilots then sided up to the jet, flying only a few meters beside its right cockpit window. The copilot was looking right at him. He then appeared to signal, indicating that the Libyan crew had understood what was going on and that they were going to comply with the interceptors. But it did not change course, nor did it descend.

At 14:01, the Israelis decided to fire highly luminescent tracer shells in front of the airliner's nose, to force it to respond. It did. The airliner descended and turned toward the Refidim Airbase. But then, when it had reached 5000 feet and lowered its landing gear, the airliner's crew seemed to change its mind. Suddenly it broke off the approach, started climbing again, putting away the landing gear, and turned west. It looked like an escape.

The Israelis were bewildered. A captain's main priority is the safety of his or her passengers: doing what this Libyan crew was doing showed none of that concern. So maybe the aircraft had been commandeered and the passengers (and crew) were along for the ride, or perhaps there were no passengers onboard at all. Still, these were only assumptions. It would be professional, the right thing to do, to double-check. The Israeli Air Force commander decided that the Phantoms should take a closer look, again.

At 14:05, one of the Phantoms flew by the airliner within a few meters and reported that all the window blinds were drawn. The Air Force commander became more and more convinced that it may have been an attempted, but foiled, terrorist attack. Letting the aircraft get away now would only leave it to have another go later.

At 14:08, he gave the order for the Israeli pilots to fire at the edges of the wings of the airliner, so as to force it to land. The order was executed. But even with the tip of its right wing hit, the airliner still did not obey the orders and continued to fly westward. The Israelis opened all international radio channels, but could not identify any communication related to this airliner. Two minutes later, the Israeli jets were ordered to fire at the base of the wings. This made the airliner descend and aim, as best it could, for a flat sandy area to land on. The landing was not successful. At 14:11, the airliner crashed and burned.

A Normative, Culpable Mistake

Had the wreckage on the ground revealed no passengers, and a crew intent on doing damage to Israeli targets, the decisions of the relevant people within the Israeli Air Force would have proven just and reasonable. There would be no basis for asserting negligence. As it turned out, however, the airliner was carrying passengers. Of 116 passengers and crew, 110 were killed in the crash.

The cockpit voice recorder revealed a completely different reality, a different "truth." There had been three crew members in the cockpit: a French captain, a Libyan copilot, and a French flight engineer (sitting behind the two pilots). The captain and the flight engineer had been having a conversation in French, while enjoying a glass of wine. The copilot evidently had no idea what they were talking about, lacking sufficient proficiency in French. It was clear that the crew had no idea that they were deviating more than 70 miles from the planned route, first flying over Egyptian and later Israeli war zones.

At 13:44, the captain first became uncertain of his position. Instead of consulting with his copilot, he checked with his flight engineer (whose station has no navigational instruments), but did not report his doubts to Cairo approach. At 13:52, he got Cairo's permission to start a descent toward Cairo International Airport. At 13:56, still uncertain about his position, the captain tried to receive Cairo's radio navigation beacon, but got directions that were contrary to those he had expected on the basis of his flight plan (as the airport was now gliding away further and further behind him).

Wanting to sort out things further, and hearing nothing else from Cairo approach, the crew continued on their present course. Then, at 13:59, Cairo came on the radio to tell the crew that they were deviating from the airway. They should "stick to beacon and report position." The Libyan copilot now reported for the first time that they were having difficulties in getting the beacon.

At 14:00, Cairo approach asked the crew to switch to Cairo control: a sign that they believed the airliner was not within range to land, close to the airport. Two minutes later the crew told Cairo control that they were having difficulties receiving another beacon (Cairo NDB, or non-directional beacon, with a certified range of about 50 kilometers), but did not say they were uncertain of their position. Cairo control asked the aircraft to descend to 4000 feet.

Not much later, the copilot reported that they had "four Mikoyan and Gurevich—Russian-built fighter airplanes (MIGs)" behind them, mistaking the Israeli Phantoms for Soviet-built Egyptian fighter jets. The captain added that he guessed they were having some problems with their heading and that they now had four MIGs behind them. He asked Cairo for help in getting a fix on his position. Cairo responded that their ground-based beacons were working normally, and that they would help find the airliner by radar.

Around that same time, one of the Phantoms had hovered next to the copilot's window. The copilot had signaled back, and turned to his fellow crewmembers to tell them. The captain and flight engineer once again engaged in French about what was going on, with the captain angrily complaining about the Phantom's signals, that this was not the way to talk to him. The copilot did not understand.

At 14:06 Cairo control advised the airliner to climb to 10,000 feet again, as they were not successful in getting a radar fix on the airplane (it was way out of their area and probably not anywhere near where they expected it to be). Cairo had two airfields: an international airport on the west side and a military airbase on the east. The crew likely interpreted the signals from the MIGs as them having overshot the Cairo international airport, and that the fighter jets had come to guide them back. This would explain why they suddenly climbed back up after approaching the Refidim airbase. Suspecting that they had lined up for Cairo East (the military field), now with fighters on their tail, the crew decided to turn west and find the international airport instead.

At 14:09, the captain snapped at Cairo control that they were "now shot by *your* fighter," upon which Cairo said they were going to tell the military that they had an unreported aircraft somewhere out there but did not know where it was. When they were shot at again, the crew panicked, accelerating their speaking in French. Were these Egyptians crazy? Then, suddenly, the copilot identified the fighters as Israeli warplanes. It was too late, with devastating consequences.

Hindsight and Culpability

The same actions and assessments that represent a conscientious discharge of professional responsibility can, with knowledge of outcome, become seen as a culpable, normative mistake.

With knowledge of outcome, we know what the commanders or pilots should have checked better (because we now know what they missed, for example, that there were passengers on board and that the jet was not hijacked). After the fact, there are always opportunities to remind professionals what they could have done better (Could you not have checked with the airline? Could your fighters not have made another few passes on either side to see the faces of passengers?). Again, had the airliner not contained passengers, nobody would have asked those questions. The professional discharge of duty would have been sufficient if that had been the outcome. And, conversely, had the Israelis known that the airliner contained passengers, and was not hijacked but simply lost, they would never have shot it down.

Few in positions to judge the culpability of a professional mistake have as much (or any) awareness of the debilitating effects of hindsight. Judicial proceedings, for example, will stress how somebody's behavior did not make sense, how it violated narrow standards of practice, rules, or laws.

Jens Rasmussen once pointed out that if we find ourselves (or a prosecutor) asking, "How could they have been so negligent, so reckless, so irresponsible?" it is not because the people in question were behaving bizarrely. It is because we have chosen the wrong frame of reference for understanding their behavior. The frame of reference for understanding people's behavior, and judging whether it made sense, is their own normal work context, the context they were embedded in. This is the point of view from where decisions and assessments are sensible, normal, daily, unremarkable, expected. The challenge, if we really want to know whether people anticipated risks correctly, is to see the world through their eyes, *without* knowledge of outcome, without knowing exactly which piece of data will turn out critical afterward.

The Worse the Outcome, the More to Account For

If an outcome is worse, then we may well believe that there is more to account for. That is probably fundamental to the social nature of accountability. We may easily believe that the consequences should be proportional to the outcome of somebody's actions. Again, this may not be seen as fair: recall the example from the beginning of this chapter. Physicians believe that liability is connected to outcomes that they have little control over, not to the quality of care they provided. To avoid liability, in other words, you don't need to invest in greater quality of care. Instead, you invest in defensive medicine: more tests, covering your back at every turn.

The main question for a just culture is not about matching consequences with outcome. It is this: Did the assessments and actions of the professionals at the time make sense, given their knowledge, their goals, their attentional demands, their organizational context? Satisfying calls for accountability here would not be a matching of bad outcome with bad consequences for the professionals involved. Instead, accountability could come in the form of reporting or disclosing how an assessment or action made sense at the time, and how changes can be implemented so that the likelihood of it turning into a mistake declines.

3 Safety Reporting and Honest Disclosure

A basic premise of a just culture is that it helps people report safety issues without fear of the consequences. It is almost an article of faith in the safety community that reporting helps learning, and that such learning helps improve safety. This seems borne out by the many incidents and accidents that, certainly in hindsight, seem to have been preceded by sentinel events. There is an implicit understanding that reporting is critical for learning. And learning is critical for improving safety (or, if anything, for staying just ahead of the changing nature of risk). The safety literature, however, has been modest in providing systematic evidence for the link among reporting, learning, and safety—if anything because we might never know the accidents that a safety report and subsequent improvements prevented from happening. For the purposes of this book, however, let us work off the premise that both honest disclosure and nonpunitive reporting have important roles to play in the creation of a safe organization. And in the creation of a just culture.

This chapter looks at safety reporting in more detail, and examines its links to a just culture inside of your organization. It also considers the regulatory and legal environment that surrounds your organization: How does that influence what you can and need to do inside? It also discusses disclosure, the obligations for it as well as the risks it creates for practitioners, and some possible protections that organizations might put in place. Chapter 4 deals in more detail with the surrounding environment, particularly the criminalization of "human error."

I once overheard a conversation between two air traffic controllers. They were talking about an incident in their control center. They discussed what they thought had happened, and who had been involved. What should they do about it?

"Remember," said one controller to the other, "Omertà!"

The other nodded, and smiled with a frown.

I said nothing but wondered silently: "Omertà"?

Surely this had something to do with the Mafia. Not with professional air traffic controllers.

Or any other professionals.

Indeed, a common definition of omertà is "code of silence." It seals people's lips. It also refers to a categorical prohibition to collaborate with authorities. These controllers were not going to talk about this incident. Not to anybody else, or anybody in a position of authority in any case. Nor were they going to collaborate voluntarily with supervisors, managers, investigators, and regulators.

I live my professional life in occasional close contact with professional groups—firefighters, pilots, nurses, physicians, police, nuclear power plant operators, inspectors, air traffic controllers. A "code of silence" is enforced and reproduced in various ways.

A senior captain with a large, respectable airline, who flies long-distance routes, told me that he does not easily volunteer information about incidents that happen on his watch. If only he and his crew know about the event, then they typically decide that that knowledge stays there. No reports are written; no "authorities" are informed.

"Why not?" I wanted to know.

"Because you get into trouble too easily," he replied. "The airline can give me no assurance that information will be safe from the prosecutor or anybody else. So I simply don't trust them with it. Just ask my colleagues. They will tell you the same thing."

I did. And they did.

Professionals under these circumstances seem to face two bad alternatives:

- Either they report a mistake and get into some kind of trouble for it (they are stigmatized, reprimanded, fired, or even prosecuted).
- Or they do not report the mistake and keep their fingers crossed that nobody else will do so either ("Remember: omertà!").

The professionals I talked to know that they can get into even worse trouble if they don't report and things come out anyway. But to not talk, and hope nobody else does either, often seems the safest bet. From the two bad alternatives, it is the less bad one.

I once spoke at a meeting at a large teaching hospital, attended by hundreds of healthcare workers. The title of the meeting was "I got reported." The rules of the country where the meeting was held say that it is the nurse's or doctor's boss who determines whether an incident should be reported to the authorities. And the boss then does the reporting. "I got reported" suggests that the doctor or nurse is at the receiving end of the decision to report: a passive nonparticipant. A casualty, perhaps, of forces greater than themselves, and interests other than their own. The nurse or doctor may have to go to his or her boss to report a mistake. But what motives do they have to do so? The formal account of what happened, and what to do about it, ultimately rests in the hands of the boss.

A FEW BAD APPLES?

We could think that professionals who rely on "omertà" are simply a few bad apples. They are uncooperative, unprofessional exceptions. Most professions, after all, carry an obligation to report mistakes and problems. Otherwise their system cannot learn and improve. So if people do not want to create safety together, there must be something wrong with them.

This, of course, would be a convenient explanation. And many rely on it. They will say that all that people need to do is report their mistakes. They have nothing to fear. Report more! Then the system will learn and get better. And you will have a part in it. Indeed, almost every profession I have worked with complains about a lack of reporting.

Yet I am not surprised that people sometimes don't want to report. The consequences can be negative. Organizations can respond to incidents in many ways. In the aftermath of an incident or accident, pressures on the organization can be severe. The media wants to know what went wrong. Politicians may too. They all want to know what the organization is going to do about it. Who made a mistake? Who should be held responsible? Even a prosecutor may become interested. National laws (especially those related to freedom of information) mean that data that people voluntarily submit about mistakes and safety problems can easily fall into wrong hands. Also, I can understand that people sometimes don't want to report because they have lost trust in the system, or in their manager, or in their organization, to do anything with their reports and the concerns in them.

Not reporting is hardly about a few bad apples. It is about structural arrangements and relationships between parties that either lay or deny the basis for trust. Trust is necessary if you want people to share their mistakes and problems with others. Trust is critical. But trust is hard to build and easy to break.

GETTING PEOPLE TO REPORT

Getting people to report can be difficult. Keeping up the reporting rate once the system is running can be equally difficult, though often for different reasons. Getting people to report is about two major things:

- Maximizing accessibility
- Minimizing anxiety

The means for reporting must be accessible. If you have reporting forms, they need to be easily and ubiquitously available, and should not be cumbersome to fill in or send up. Computer-based systems, of course, can be made as user-friendly or unfriendly as the fantasy of their developer allows. But what about anxiety? Initially, people will ask questions like

- What will happen to the report?
- Who else will see it?
- Do I jeopardize myself, my career, my colleagues?
- Does this make legal action against me easier?

You should ask yourself whether there is a written policy that explains to everybody in the organization what the reporting process looks like; what the consequences of reporting could be; and what rights, privileges, protections, and obligations people may expect. Without a written policy, ambiguity can persist. And ambiguity means that people will be less inclined to share safety-critical or risk-related information with you.

Getting people to report is about building trust: trust that the information provided in good faith will not be used against those who reported it. Such trust must be built in various ways. An important way is by structural (legal) arrangement. Making sure people have knowledge about the organizational and legal arrangements surrounding reporting is very important: disinclination to report is often related more to uncertainty about what *can* happen with a report than by any real fear about what *will* happen. One organization, for example, has handed out little credit-sized cards to its employees to inform them about their rights and duties around an incident.

Another way to build trust is by historical precedent: making sure there is a good record for people to lean on when considering whether to report an event or not. But as mentioned previously, trust is hard to build and easy to break: one organizational or legal response to a reported event that shows that divulged information can somehow be used against the reporter can destroy months or years of building goodwill.

WHAT TO REPORT?

The belief is that reporting contributes to organizational learning. It is to help prevent recurrence by making systemic changes that aim to redress some of the basic circumstances that went awry. This means that any event that has the potential to shed some light on (and help improve the conditions for safe practice) is, in principle, worth reporting and investigating. But that still does not create very meaningful guidance.

Many professions have codified the obligation to report. The Eurocontrol Safety and Regulatory Requirement (ESARR 2), for example, told air traffic controllers and their organizations that "all safety occurrences need to be reported and assessed, all relevant data collected and lessons disseminated." Saying that all safety occurrences need to be reported is easy. But what counts as a "safety occurrence?" This can be open for interpretation: the missed approach of the 747 in the case study about honest mistakes was, according to the pilot, not a safety occurrence. It was not worth reporting. But according to his bosses and regulators, it was. And the fact that he did not report it made it all the more so.

Professional codes about reporting, then, should ideally be more specific than saying that "all safety occurrences" should be reported. What counts as a clear opportunity for organizational learning for one, perhaps constitutes a dull and not report worthy event to somebody else. Something that could have gone terribly wrong, but did not, is not necessarily a clear indication of reportworthiness either. After all, in many professions things can go terribly wrong the whole time ("I endanger my passengers every day I fly!"). But that does not make reporting everything particularly meaningful.

Which event is worthy of reporting and investigating is, at its heart, a judgment. First, it is a judgment by those who perform safety-critical work at the sharp end. Their judgment about whether to report something is shaped foremost by experience—the ability to deploy years of practice into gauging the reasons and seriousness behind a mistake or adverse event.

To be sure, those years of experience can also have a way of blunting the judgment of what to report. If all has been seen before, why still report? What individuals and groups define as "normal" can glide, incorporating more and more nonconformity as time goes by and as experience mounts. In addition, the rhetoric used to talk about mistakes can serve to "normalize" (or at least deflect) an event away from the professionals at that moment. A "complication" or "noncompliant patient" is not so compelling to report (though perhaps worth sharing with peers in some other forum), as when the same event were to be denoted as, for example, a diagnostic error.

Whether an event is worth reporting, in other words, can depend on what language is used to describe that event in the first instance. This has another interesting implication: In some cases a lack of experience (either because of a lack of seniority or inexperience with a particular case, or in that particular department) can be immensely refreshing in questioning what is "normal" (and thus what should be reported or not).

Investing in a meeting where different stakeholders share their examples of what is worth reporting could be useful. It could result in a list of examples that can be handed to people as partial guidance on what to report. But in the end, given the uncertainties about how things can be seen as valuable by other people, and how they could have developed, the ethical obligation might well be: "If in doubt, report."

But then, what delimits an "event?" The reporter needs to decide where the reported event begins and ends. She or he needs to decide how to describe the roles and actions of other participants who contributed to the event (and to what extent to identify other participants, if at all). Finally, the reporter needs to settle on a description that offers the organization a chance to understand the event and find leverage for change. Many of these things can be structured beforehand, for example, by offering a reporting form that gives guidance and asks particular questions ("need-to-know" for the organization to make any sense of the event) as well as ample space for free-text description.

KEEPING THE REPORTS COMING IN

Keeping up the reporting rate is also about trust. But it is even more about involvement, participation, and empowerment. Building enough trust so that people do not feel put off to send in reports in the first place is one thing. Building a relationship with participation and involvement that will actually get and keep people to send in reports is quite another.

Many people come to work with a genuine concern for the safety and quality of their professional practice. If, through reporting, they have an opportunity to actually contribute to visible improvements, then few other motivations or exhortations to report are necessary. Making a reporter part of the change process can be a good way forward, but this implies that the reporter wants (or dares) to be identified as such, and that managers have no problems with integrating employees in their work for improved safety and quality.

Sending feedback into the department about any changes that result from reporting can also be a good strategy. But it should not become the stand-in for doing anything else with the reports. Many organizations get captured by the belief that reporting is a virtue in itself: if only people report mistakes, and their self-confessions are distributed back to the operational community, then things will automatically improve and people will feel motivated to keep reporting. This does not work for long. Active engagement with that which is reported, and perhaps even with those who report, is necessary. Active, demonstrable intervention that acts on the reported information is too.

REPORTING TO MANAGERS OR TO SAFETY STAFF?

In many organizations, the line manager is the recipient of reports. This makes (some) sense: the line manager probably has responsibility for safety and quality in the primary processes, and should have the latest information on what is or is not going well. But this practice has some side effects.

- One is that it hardly renders reporters anonymous (given the typical size of a department), even if no name is attached to the report.
- The other is that reporting can have immediate line consequences (an unhappy manager, consequences for one's own chances to progress in career).
- Especially in cases where the line manager her- or himself is part of the problem the reporter wishes to identify, such reporting arrangements all but stop the flow of useful information.

I remember studying one organization that had shifted from a reporting system run by line managers to one run by safety-quality staff. Before the transition, employees actually turned out very ready to confess an "error" or "violation" to their line manager. It was almost seen as an act of honor. Reporting it to a line organization—which would see an admission of error as a satisfactory conclusion to its incident investigation—produced rapid closure for all involved. Management would not have to probe deeper, as the operator had seen the error of his or her ways and had been reprimanded and told or trained to watch out better next time.

For the operators, simply and quickly admitting an error avoided even more or deeper questions from their line managers. Moreover, it could help avert career consequences, in part by preventing information from being passed to other agencies (e.g., the industry's regulator). Fear of retribution, in other words, did not necessarily discourage reporting. In fact, it encouraged a particular kind of reporting: a mea culpa with minimal disclosure that would get it over with quickly for everybody. "Human error" as cause seemed to benefit everyone—except organizational learning.

As one employee told us: "I didn't tell the truth about what took place, and this was encouraged by the line manager. He had made an assumption that the incident was due to one factor, which was not the case. This helped me construct and maintain a version of the story that was more favorable for us (the frontline employees)."

Perhaps the most important reason to consider a reporting system that is not run just by line management is that it can radically improve organizational learning. Here is what one line manager commented after having been given a report by an operator.

"The incident has been discussed with the operator concerned, pointing out that priorities have to be set according to their urgency. The operator should not be distracted by a single problem and neglect the rest of his working environment. He has been reminded of applicable rules and allowable exceptions to them. The investigation report has been made available to other operators by posting it on the internal safety board."

Such countermeasures really do not represent the best in organizational learning. In fact, they sound like easy feel-good fixes that are ultimately illusory. Or simply very short-lived. Opening up a parallel system (or an alternative one) can help. The reports in this system should go to a staff officer, not a line manager (e.g., a safety or quality official), who has no stakes in running the department. The difference between what gets reported to a line manager and that is written in confidential reports can be significant. Also, the difference in understanding, involvement, and empowerment that the reporter feels can be significant.

"It is very good that a colleague, who understands the job, performs the interviews. They asked me really useful questions and pointed me in directions that I hadn't noticed. It was very positive compared to before. Earlier you never had the chance to understand what went wrong. You only got a conclusion to the incident. Now it is very good that the report is not published before we have had the chance give our feedback. You are very involved in the process now and you have time to go through the occurrence. Before you were placed in the hot chair and you felt guilty. Now, during interviews with the safety staff, I never had the feeling that I was accused of anything."

Of course, keeping a line-reporting mechanism in place *can* be productive for continuous improvement work, especially if things need to be brought to the attention of relevant managers immediately. But you should perhaps consider a separate, parallel confidential reporting system if you don't already have one. Line-based and staff-based (or formal and confidential) reporting mechanisms offer different kinds of leverage for change. Not mining both data sources for information could mean your organization is losing out on improvement opportunities.[30]

THE SUCCESSFUL REPORTING SYSTEM: VOLUNTARY, NONPUNITIVE, AND PROTECTED

In summary, near-miss reporting systems that work in practice do a few things really well.[63] They are

- Voluntary
- Nonpunitive
- Protected

Voluntary

The language in the Eurocontrol rules suggests that reporting should be compulsory. All safety occurrences *need* to be reported. Many organizations have taken it to mean just that, and have told their employees that they are obligated to report safety occurrences. But this might actually make little sense. Why does research show that a voluntary system is better, leaving it to people's own judgment to report or not? Mandatory reporting implies that everybody has the same definition of what is risky, what is worthy of reporting. This, of course, is seldom the case. The airline captain who maintained omertà on his own flight deck could—in principle—be compelled to report *all* incidents and safety occurrences, as Eurocontrol would suggest. But the captain probably has different ideas about what is risky and worth reporting than his organization does.

So who gets to say what is worthy of reporting? If the organization claims that right, then they need to specify what they expect their professionals to report. This turns out to be a pretty hopeless affair. Such guidance is either too specific or too general. All safety occurrences, like Eurocontrol suggests? Sure, then of course practitioners will decide that what they were involved in wasn't really a safety occurrence. Well, the organization may come back and put all kinds of numeric borders in place. If it is lower than 1000 feet, or closer than 5 miles or longer than 10 minutes, then it is a safety occurrence. But a line means a division, a dichotomy. Things fall on either side. What if a really interesting story unfolds just on the other side, the safe side of that border—at 9 minutes and 50 seconds? Or 1100 feet? Or 5.1 miles? Of course, you might say, we have to leave that to the judgment of the professional. Ah! That means that the organization does not decide what should be reported! It is the professional again. That is where the judgment resides. Which means it is voluntary (at least on the safe side of those numbers), even though you think it is compulsory.

The other option is to generate a very specific list of events that will need to be reported. The problem there is that practitioners can then decide that on some little nuance or deviation their event does not match any of the events in the list. The guidance will have become *too* specific to be of any value. Again, the judgment is left up to the practitioner, and the system once again becomes voluntary rather than compulsory.

Also, making reporting mandatory implies some kind of sanction if something is not reported, which destroys the second ingredient for success: having a nonpunitive system. It may well lead to a situation in which practitioners will get smarter at making evidence of safety-critical events go away (so they will not get punished for not reporting them). As said previously, practitioners can engage in various kinds of rhetoric or interpretive work to decide that the event they were involved in was not a near miss, not an event worthy of reporting or taking any further action on.[64] Paradoxically then, a mandatory system can increase underreporting—simply because the gap between what the organization expects to be reported and what gets reported gets stretched to its maximum.

Nonpunitive

Nonpunitive means that the reporter is not punished for revealing his or her own violations or other breaches or problems of conduct that might be construed as culpable.

In other words, if people report their honest mistakes in a just culture, they will not be blamed for them. The reason is that an organization can benefit much more by learning from the mistakes that were made than from blaming the people who made them. So people should feel free to report their honest mistakes.

The problem is, often they don't.

Often they don't feel free, and they don't report.

This is because reporting can be risky. Many things can be unclear:

- How exactly will the supervisor, the manager, or the organization respond?
- What are the rights and obligations of the reporter?
- Will the reported information stay inside of the organization? Or will other parties (media, prosecutor) have access to it as well?

The reason why people fail to report is not because they want to be dishonest. Nor because they are dishonest. The reason is that they fear the consequences. And often with good reason.

- Either people simply don't know the consequences of reporting, so they fear the unknown, the uncertainty
- Or the consequences of reporting really can be bad, and people fear invoking such consequences when they report information themselves

Although the first reason may be more common, either reason means that there is serious work to do for your organization. In the first case, that work entails clarification. Make clear what the procedures and rules for reporting are, what people's rights and obligations are, and what they can expect in terms of protection when they report. In the second case it means trying to make different structural arrangements, for example, with regulators or prosecutors, with supervisors or managers, about how to treat those who report. This is much more difficult, as it involves the meshing of several different interests.

Not punishing that which gets reported makes great sense, simply because otherwise it won't get reported. This of course creates the dilemma for those receiving the report (even if via some other party, e.g., a quality and safety staff): they want to hear everything that goes on but cannot accept that it goes on. The willingness to report anything by any other nurse would have taken a severe beating. The superficially attractive option is to tell practitioners (as much guidance material around reporting suggests) that their reports are safe in the hands of their organization unless there is evidence of bad things (such as negligence or deliberate violations). Again, such guidance is based on the illusion of a clear line between what is acceptable and what is not—as if such things can be specified generically, beforehand. They can't.

From the position of a manager or administrator, one way to manage this balance is to involve the practitioners who would potentially report (not necessarily the one who *did* report, because if it's a good reporting system, that might not be known to the manager; see later). What is their assessment of the event that was reported? How would they want to be dealt with if it were their incident, not their colleague's? Remember that perceived justice lies less in the eventual decision than in who and what is involved in making that decision. For a manager, keeping the dialogue open with her or his practitioner constituency must be the most important aim. If dialogue is killed by rapid punitive action, then a version of the dialogue will surely continue elsewhere (behind the back of the manager). That leaves the organization none the wiser about what goes on and what should be learned.

Protected

Finally, successful reporting systems are *protected.* That means that reports are confidential rather than anonymous. What is the difference? "Anonymity" typically means that the reporter is never known, not to anybody. No name or affiliation has to be filled in anywhere. "Confidentiality" means that the reporter fills in name and affiliation and is thus known to whomever gets the report. But from there on, the identity is protected, under any variety of industrial or organizational or legal arrangements. If reporting is anonymous, two things might happen quickly. The first is that the reporting system becomes the trash can for any kind of vitriol that practitioners may accumulate about their job; their colleagues; or their hospital during a workday, workweek, or career. The risk for this, of course, is larger when there are few meaningful or effective line management structures in place that could take care of such concerns and complaints. However, senseless and useless bickering could clog the pipeline of safety-critical information. Signals of potential danger would get lost in the noise of grumble. So that's why confidentiality makes more sense. The reporter may feel some visibility, some accountability even, for reporting things that can help the organization learn and grow.

A second problem with an anonymous reporting system is that the reporter cannot be contacted if the need arises for any clarifications. The reporter is also out of reach for any direct feedback about actions taken in response to the report. The NASA Aviation Safety Reporting System (ASRS) recognized this quickly after finding reports that were incomplete or could have been much more potent in revealing possible danger if only this or that detail could be cleared up. As soon as a report is received, the narrative is separated from any identifying information (about the reporter and the place and time of the incident) so that the story can start to live its own life without the liability of recognition and sanction appended to it. This recipe has been hugely successful. ASRS receives more than 1000 reports each week.[65] Of course, gathering data is not the same as analyzing it, let alone learning from it. Indeed, such reporting systems can become the victims of their own success: the more data you get, the more difficult it can be to make sense of it all, certainly if the gathering of data outpaces the analytic resources available to you.

What if Reported Information Falls into the Wrong Hands?

Most democracies have strong freedom-of-information legislation. This allows all citizens from the outside access, in principle, to nonconfidential information. Such transparency is critical to democracy, and in some countries, freedom-of-information is even enshrined in the constitution. But the citizen requesting information can easily be an investigating journalist, a police investigator, or a prosecutor. Freedom-of-information is really an issue when the organization itself is government-owned (as many hospitals or air traffic service providers or even some airlines are). Moreover, safety investigating bodies are also government organizations and thus subject to such legislation. This can make people unwilling to collaborate with safety investigators.

The potential for such exposure can create enormous uncertainty. And uncertainty typically dampens the willingness of people to report. People become anxious to leave information in files with their organization. In fact, the organization itself can become anxious to even have such files. Having them creates the risk that names of professionals end up in the public domain. This in turn can subject safety information to oversimplification and distortion and misuse by those who do not understand the subtleties and nuances of the profession.

THE DIFFERENCE BETWEEN DISCLOSURE AND REPORTING

For the purposes of just culture, it is useful to make a distinction between reporting and disclosure (Table 3.1)[1]:

- **Reporting is the provision of information to supervisors, oversight bodies, or other agencies**. Reporting means given a spoken or written account of something that you have observed, participated in, or done to an appointed party (supervisor, safety manager). Reporting is thought necessary because it contributes to organizational learning. Reporting is not primarily about helping customers or patients, but about helping the organization (e.g., colleagues) understand what went wrong and how to prevent recurrence.
- **Disclosure is the provision of information to customers, clients, patients, and families**. The ethical obligation to disclose your role in adverse events comes from a unique, trust-based relationship with the ones who rely on you for a product or service. Disclosure can be seen as a marker of professionalism. Disclosure means making information known, especially information that was secret or that could be kept secret. Information about incidents that only one or a few people were involved in, or that only professionals with inside knowledge can really understand, could qualify as such.

TABLE 3.1
The Difference between Disclosure and Reporting for Individuals and Organizations

	Reporting	Disclosure
Individual	Providing a written or spoken account about observation or action to supervisors, managers, safety/quality staff	Making information known to customers, clients, patients
Organization	Providing information about employees' actions to regulatory or other (e.g., judiciary) authorities when required	Providing information to customers, clients, patients, or others affected by the organization's or employees' actions

As with any complex problem, neither reporting nor disclosure is a constant guarantee for a just culture. Reporting, for example, can be used (like in the preceding examples) as a way to deny responsibility and engagement with the problem. When certain protections are in place, reporting can perhaps even be used to insulate oneself from accountability. Disclosure, too, can sometimes lead to injustice. Disclosure *beforehand* to clients or patients or customers can be used to immunize an organization (or individuals in it) from legal or other recourse after something has gone wrong. The spread of subprime mortgages in the 2000s, for example, led to complex transactions with clients to whom everything was disclosed that the law asked for—though in inscrutable fine print. After the bubble burst, and people started losing homes, banks fended off lawsuits by arguing that they had done and disclosed everything the law required, and that people had confirmed with their signature that they understood it all. That can be the harm and injustice of disclosure.

- **Practitioners typically have an obligation to report** to their organization when something went wrong. As part of the profession, they have a duty to flag problems and mistakes. After all, they represent the leading edge, the sharp end of the system: they are in daily contact with the risky technology or business. Their experiences are critical to learning and continuous improvement of the organization and its work.
- **Many practitioners also have an obligation to disclose** information about things that went wrong to their customers, clients, and patients. This obligation stems from the relationship of trust that professionals have with those who make use of their services.
- **Organizations also have an obligation to disclose** information about things that went wrong. This obligation stems from the (perhaps implicit) agreement that companies have with those who make use of their services or are otherwise affected by their actions.
- **Organizations (employers), the judiciary, and regulators have an obligation to be honest about the possible consequences** of failure, so that professionals are not left in the dark about what can happen to them when they do report or disclose.
- One could also propose that **organizations have a (legal) obligation to report** certain things to other authorities (judiciary, regulatory).

Disclosure and reporting can clash. And different kinds of reporting can also clash. This can create serious ethical dilemmas that both individual professionals and their employing organizations need to think about.

- If an organization wants to encourage reporting, it may actually have to curtail disclosure. Reporters will step forward with information about honest mistakes only when they feel they have adequate protection against that information being misused or used against them. This can mean that reported information must somehow remain confidential, which rules out disclosure (at least of that exact information).

- Conversely, disclosure by individuals may lead to legal or other adverse actions (even against the organization), which in turn can dampen people's or the organization's willingness to either report or disclose.
- If organizations report about individual actions to regulatory or judicial authorities, this too can lower the willingness to report (and perhaps even disclose) by individuals, as they feel exposed to unjust or unwelcome responses to events they have been involved in.

A representative of the regulator had been sent out for a field visit as a customer of an organization I once worked with. She had observed things in the performance of one practitioner that, according to the rules and regulations, weren't right. Afterwards, she contacted the relevant managers in the organization and let them know what she had seen. The managers, in turn, sent a severe reprimand to the professional.

It really strained trust and the relationship between practitioners and management: by their reaction, it was clear that reporting was not encouraged. The regulator would not have been happy either to find out that her visit was being used as a trigger to admonish an individual practitioner instead of resolving more systemic problems. It was as if the managers were offloading their responsibility for the problems observed onto the individual practitioner.

Overlapping Obligations

The difference between disclosure and reporting is not as obvious or problematic in all professions.

- Where individual professional contact with clients is very close, such as in medicine, reporting and disclosure are two very different things.
- Where the relationship is more distant, such as in air traffic control, the distinction blurs because for individual air traffic controllers there is not immediately a party to disclose to. The air traffic control organization, however, can be said to have an obligation to disclose.

If organizational disclosure or reporting does not occur, then the mistakes made by people inside that organization may no longer be seen as honest, and the organization can get in trouble as a result. This goes for individuals, too. It may have played a role in the case in Chapter 2, as it plays a role in many other cases. Not providing an account of what happened may give other people the impression that there is something to hide. And if there is something to hide, then what happened may not be seen as an "honest" mistake.

The killing of a British soldier in Iraq by a US pilot was a "criminal, unlawful act," tantamount to manslaughter, a British coroner ruled. The family of Lance Corporal Hull, who died in March 2003, was told at the inquest in Oxford, England, that it was "an entirely avoidable tragedy." His widow, Susan, welcomed the verdict, saying it was what the family had been waiting four years for. Hull said she did not want to see the pilot prosecuted, but felt she been "badly let down" by the US government, which consistently refused to cooperate.

Susan Hull had also been told by the UK Ministry of Defense that no cockpit tape of the incident existed. This was proven untrue when a newspaper published the tape's contents and when it was later posted on the Internet. It showed how Hull was killed when a convoy of British Household Cavalry vehicles got strafed by two US A10 jets. The British Ministry of Defense issued an apology over its handling of the cockpit video, while the US Department of Defense denied there had been a cover-up and remained adamant that the killing was an accident.

The coroner, Andrew Walker, was damning in his appraisal of the way the Hull family had been treated. "They, despite request after request, have been, as this court has been, denied access to evidence that would provide the fullest explanation to help understand the sequence of events that led to and caused the tragic loss of LCorp Hull's life," he said. "I have no doubt of how much pain and suffering they have been put through during this inquisition process and to my mind that is inexcusable," he said.[66]

Nondisclosure in the wake of an incident often means that a mistake will no longer be seen as honest. And once a mistake is considered dishonest, people may no longer care as much about what happens to the person who made that mistake, or to the party (e.g., the organization) responsible for withholding the information. This is where a mistake can get really costly—both financially and in terms of unfavorable media exposure, loss of trust and credibility, regulatory scrutiny, or even legal action.

I recall one adverse event in which the family was very upset, not only about what had happened, but also about the organization not being seen as forthcoming. The family had been invited by the organization to stay in a nice hotel for some of the legal proceedings. Feeling injured and let down, they ordered as much expensive room service as possible and then threw it all away. Having been hurt by the organization, they wanted to hurt the organization as much as possible in return.

Nondisclosure is often counterproductive and expensive. Silence can be interpreted as stone-walling, as evidence of "guilty knowledge." It is well known that lawsuits in healthcare are often more a tool for discovery than a mechanism for making money. People don't generally sue (in fact, very few actually do). But when they do, it is almost always because all other options to find out what happened have been exhausted.[17] Paradoxically, however, lawsuits still do not guarantee that people will ever get to know the events surrounding a mishap. In fact, once a case goes to court, "truth" will likely be the first to suffer. The various parties may likely retreat into defensive positions from which they will offer only those accounts that offer them the greatest possible protection against the legal fallout.

THE RISKS OF REPORTING AND DISCLOSURE

John Goglia, a former member of the National Transportation Safety Board, recently wrote how Southwest Airlines settled a whistleblower lawsuit.[67] *It was filed by a mechanic who said he was disciplined for finding and reporting two cracks in the fuselage of a Boeing 737-700 while performing a routine maintenance check.*

Southwest Airlines agreed to remove the disciplinary action from the mechanic's file and to pay him $35,000 in legal fees. The lawsuit was filed under the whistleblower protections, whose statute provides an appeal process for airline workers who are fired or otherwise disciplined for reporting safety information. The settlement was reached after a January 8, 2015 Department of Labor Administrative Judge dismissed Southwest's motion for summary judgment and granted in part the mechanic's motion for summary judgment.

The judge's decision summarizes the allegations as follows: "On the evening of July 2, 2014, the [mechanic] was assigned by [Southwest] to perform a [maintenance] check on a Southwest Boeing 737-700 aircraft, N208WN. This maintenance check is part of Southwest's Maintenance Procedural Manual (MPM). This check requires a mechanic to follow a task card which details the tasks to be accomplished. The task card requires the mechanic to walk around the aircraft to visually inspect the fuselage. During his inspection, the [mechanic] discovered two cracks on the aircraft's fuselage and documented them. Discovery of these cracks resulted in the aircraft being removed from service to be repaired. Thereafter, the mechanic was called into a meeting with his supervisors to discuss the issue of working outside the scope of his assigned task. He was then issued a 'Letter of Instruction' advising the mechanic that he had acted outside the scope of work in the task card and warning him that further violations could result in further disciplinary actions. The mechanic alleged in his whistleblower complaint that the letter from Southwest was calculated to, or had the effect of, intimidating [him] and dissuading him and other Southwest [mechanics] from reporting the discovery of cracks, abnormalities or defects out of fear of being disciplined."

Southwest responded to the mechanic's allegations claiming that the mechanic went outside the scope of his duties when he observed the cracks and reported them. The airline further claimed that its Letter of Instruction was issued because the mechanic worked "outside the scope of his task" and not because he reported a safety problem. It further claimed that the letter was not a disciplinary action and the mechanic was not entitled to whistleblower protection. The administrative judge sided with the mechanic in dismissing Southwest's claims and finding that the mechanic engaged in activities protected by the whistleblowing statute and that Southwest was aware of it. Although no final decision was reached on the merits of the mechanic's case, the settlement followed close on the heels of the judge's decision.

The Ethical Obligation to Report or Disclose

Being honest in reporting a safety issue, as in the preceding case, can lead to problems in the relationship with the employer. But what about not being honest? What about not acknowledging or apologizing for a mistake? This often causes relationships to break down, too—more so than the mistake or mishap itself. This makes honesty all the more important in cases where there is a prior professional relationship, such as that of a patient and doctor. In a specific example, the Code of Medical Ethics (Ethical Opinions E-8.12) of the American Medical Association has said since 1981 that

> It is a fundamental requirement that a physician should at all times deal honestly and openly with patients... Situations occasionally occur in which a patient suffers significant medical complications that may have resulted from the physician's mistake or judgment. In these situations, the physician is ethically required to inform the patient of all the facts necessary to ensure understanding of what has occurred... Concern regarding legal liability which might result following truthful disclosure should not affect the physician's honesty with a patient.

This is unique, because few codes exist that specifically tell professionals to be honest. It also spells out in greater detail which situations ("that may have resulted from a physician's mistake or judgment") are likely to fall under the provisions of the Code. So this is a good start. But it still leaves a large problem. What does "honesty" mean? Being honest means telling the truth. But telling the truth can be reduced to "not lying." If it is, then there is still a long distance to full disclosure. What to say, how much to say, or how to say it, often hinges more on the risks that people see with disclosure than with what a code or policy tells them to do.

The Risk with Disclosure

If structural arrangements and relationships inside an industry, or a profession, are such that all bets are off when you tell your story, then people will find ways to not disclose, or to disclose only in ways that protect them against the vagaries and vicissitudes of the system. Another reason may be that practitioners are simply not well prepared to disclose. There might be no meaningful training of practitioners in how to disclose and punch through the pain, shame, guilt, and embarrassment of an incident.

> Paralyzed by shame or lacking their own understanding of why the error occurred, physicians may find a bedside conversation too awkward. They may also be unwilling or unable to talk to anyone about the event, inhibiting both their learning and the likelihood of achieving resolution. Such avoidance and silence compound the harm.[33]

Programs in various countries are fortunately teaching practitioners about open disclosure, often through role play. These are generally thought to have good results.[68] It is a good way to complement or reduce the influence of a profession's "hidden curriculum." This hidden curriculum can be seen as a sort of spontaneous system of informal mentoring and apprenticeship that springs up in parallel to any formal program. It teaches, mainly through example, students and residents how to think and talk about their own mistakes and those of their colleagues. They learn, for example, how to describe mistakes so that they are no longer mistakes. Instead, they can become

- "Complications"
- The result of a "noncompliant" patient
- A "significantly avoidable accident"
- An "inevitable occasional untoward event"
- An "unfortunate complication of a usually benign procedure"

Medical professionals, in the hidden curriculum, also learn how to talk about the mistakes among themselves, while reserving another version for the patient and family and others outside their immediate circle of professionals. A successful story about a mistake is one that not only (sort of) satisfies the patient and family, but also one that protects against disciplinary measures and litigation.

Many, if not all, professions, have a hidden curriculum. Perhaps it teaches professionals the rhetoric to make a mistake into something that no longer is a mistake. Perhaps it teaches them that there is a code of silence, an omertà, that proscribes collaborating truthfully with authorities or other outside parties.

The Protection of Disclosure

The protection of disclosure should first and foremost come from structural arrangements made by the organization or profession. One form of protecting disclosure is that of "I'm sorry laws." According to these laws (now implemented in, for example, the US states of Oregon and Colorado), doctors can say to patients that they are sorry for the mistake(s) they committed. This does not offer them immunity from lawsuits or prosecution, but it does protect the apology statement from being used as evidence in court. It also does not prevent negative consequences, but at least that which was disclosed cannot be used directly against the professional. Such protection is not uncontroversial, of course. If you make a mistake, you should not only own up to it but also face the consequences, some would say. Which other professions have such cozy protections? This is where ethical principles can start to diverge.

WHAT IS BEING HONEST?

The question that comes up is this: Is honesty a goal in itself? Perhaps honesty should fulfill the larger goals of

- Learning from a mistake to improve safety
- Achieving justice in its aftermath

These are two goals that serve the common good. Supposedly pure honesty can sometimes weaken that common good. Should we always pursue honesty, or truth-telling, because it is the "right" thing to do, no matter what the consequences could be?

Dietrich Bonhoeffer, writing from his cell in the Tegel prison in Berlin in 1943 Nazi Germany, drafted a powerful essay on this question. He was being held in part on suspicion of a plot to overthrow Hitler, a plot in which he and his family were actually deeply involved.[17] If he were to tell the truth, he would have let murderers into his family. If he were to tell the truth, he would have to disclose where other conspirators were hidden. So would not telling this make him a liar? Did it make him, in the face of Nazi demands and extortions, immoral, unethical? Bonhoeffer engaged in nondisclosure, and outright deception.

The circumstances surrounding truth-telling in professions today is not likely as desperate and grave as Bonhoeffer's (he was executed in a concentration camp just before the end of the war, in April 1945). But his thoughts have a vague reflection in the fears of those who consider disclosing or reporting today. What if they tell the whole truth—rather than a version that keeps the system happy and them protected? Bonhoeffer makes a distinction between the morality and epistemology of truth-telling that may offer some help here.

- The **epistemology** of truth-telling refers to the validity or scope of the knowledge offered. In that sense, Bonhoeffer did not tell the truth (but perhaps the nurse in the Xylocard case that follows did).
- The **morality** of truth-telling refers to the correct appreciation of the real situation in which that truth is demanded. The more complex that situation, the more troublesome the issue of truth-telling becomes (the nurse in the Xylocard case may not have done this, but perhaps should have).

Bonhoeffer's goal in not disclosing was not self-preservation, but the protection of the conspiracy's efforts to jam the Nazi death machine, thereby honoring the perspective of the most vulnerable. That his tormentors wanted to know the truth was unethical, much more so than Bonhoeffer's concealment of it.

Translate this into some of the case studies that you find throughout this book. It may be less ethical for prosecutors or judges in positions of power to demand the full truth than it would have been for these professionals to offer only a version of that truth.

As we will see in Chapter 4, going to court, and demanding honesty there, becomes a different issue altogether. Once adversarial positions are lined up against each other in a trial, where one party has the capacity to wreak devastating consequences onto another, the ethics of honesty get a whole new dynamic. Demanding honesty in these cases can end up serving only very narrow interests, such as the preservation of a company's or hospital's reputation, or *their* protection from judicial pressure. Or it deflects responsibility from the regulator (particularly in countries where they employ the prosecutor) after allowing a company to routinely hand out dispensations from existing rules. Wringing honesty out of people in vulnerable positions is neither just nor safe. It does not bring out a story that serves the dual goal of satisfying calls for accountability and helping with learning. It really cannot contribute to just culture. Let's look at a case study that brings this out quite wrenchingly.

CASE STUDY

A Nurse's Error Became a Crime

Let me call the nurse Mara.

It was on a Friday in March that I first met her. I had no idea what she would look like—an ICU nurse in her late 40s, out of uniform. This could be anybody.

As I bounded up the stairs, away from the train platform, and swept around the corner of the overpass, there she was. It had to be her. Late 40s, an intensive care nurse of 25 years, a wife, a mother of three.

But now a criminal convict. An outcast. A black sheep. On sick leave, perversely with her license to practice still in the pocket.

We exchanged a glance and then embraced.

What else was I to do, to say? The telephone conversation from the night before fresh in my mind, here she was for real. Convicted twice of manslaughter in the medication death of a three-month-old girl. Walking free now, as her case was pending before the Supreme Court.

I stepped back and offered: "This sucks, doesn't it?"

She nodded, eyes glistening.

It was her all right. There hadn't been many other people around in any case.

"I recognized you from a video of a lecture you held," she explained as we turned to go down the stairs to meet her lawyer.

"And how kind of you to travel up all this way."

"Well, it's the least I could do," I said.

Snow was everywhere. Unyielding, huge piles, blanketing the little town. The lawyer's address was distinguished. The most prominent address in the town, in fact. An imposing building in stately surroundings, spacious offices, high ceilings, the quiet reverence and smell of an old library, archaic dress, archaic language.

The lawyer prattled on, clearly proud that he had, once again, scored the Big One: a hearing in the Supreme Court. I don't know whether proud lawyers make good lawyers. What I wanted to know was the planned substance of the defense, as assisting with that was my only card. A lot of the banter was inconsequential to me, much of it incomprehensible to Mara—a foreign language.

As I looked over to where Mara sat, I could not help but find her so out of place. A fish on the shore, gasping, trying to make sense of its surroundings as the burden of a final crawl for survival started sinking in. How on earth could a normal, diligent nurse who had practiced her entire adult life, ever have expected to become the lead character in somebody's lofty law offices for a prelude to an appearance at the nation's highest court? It must have felt surreal to her. She certainly looked it.

As it turned out (how naïve I am), there is no substance to speak of in a defense before the Supreme Court, because it's all form. Mara began to discover this too, haltingly, stumblingly, and increasingly disgusted.

"All I want is the truth to come out," she repeated.

"It won't," the lawyer found himself explaining over and over. "This is not about truth. It's about procedure and legal interpretation, and whether it has been correctly followed and applied. All we want is to get you off the hook. What we have to show is that the course of justice so far has been improper—the truth is secondary."

"But what about all the other people involved?" Mara appeared to become anguished. "The pediatrician, the prescription that magically disappeared days after the death, the nurse who administered the medication, the doctors who didn't really diagnose, the lousy routines at the hospital, what about them? The truth is that they are all part of this too!"

The lawyer turned to ice. "They are not on trial now, are they? This is about you. You are the only one. As soon as we bring them up in the Supreme Court, they will ask me, 'So where are those co-defendants then, counselor? We thought this case was about the nurse, not all these others.' So don't bring it up, I plead with you; don't bring it up."

Mara seemed exasperated. If justice was like this, disinterested in truth, directed through dogmatic decisions by outsiders that limited what was relevant from the events that got her here, with people putatively helping her by telling her not to argue for her case, then why bother at all? Was it worth it? Justice was supposed to be about getting out the real story, what really happened. That would be just. Justice would be about righting what was wrong, and about preventing it from happening again. That would be just too. Yes, she made a mistake; yes, a baby died. She knew that. But she also knew that the entire system in which she worked was rotten, porous, and ready to kill again.

But it was plain to me that Mara knew why she was here. It wasn't just because of her, because of her role, because of her fate, or because everybody was suddenly gathering around invigorated efforts to make healthcare safer.

She knew who was paying her lawyer, and it wasn't she. Fewer than 1% of cases presented actually get heard by the Supreme Court in my adopted country, and hers was among them. It must have mattered, somehow. The country had taken interest. The union certainly had too. Should medical practitioners involved in a patient's death be subject to the criminal justice system? Or should they be dealt with through the established professional channels: the medical disciplinary board? A great deal was at stake; that much was obvious to Mara. Realizing that she may have used the stronger solution of the medicine, she had volunteered her possible contribution to the baby's death to her boss a few days after it had happened. Her boss duly reported the event to the relevant agency, but somebody also leaked it to the local press. Mara never found out who. Her role was played up, and a prosecutor happened to read it in the morning paper. After months of uncertainty—Mara even called up the prosecutor herself one day to get clarity about her intentions—charges were brought. A local court found her guilty of manslaughter. The conviction was upheld by a higher court, which toughened the punishment. Now the case was headed for the Supreme Court. Would people in healthcare ever volunteer information about incidents again? Was this the death knell for nascent medical event reporting systems? Was patient safety going to be dealt a serious setback?

We wandered back into town, in search of a cup of coffee.

When we had slipped into the warmth of a bakery, shaken the snow off our shoulders, and sat down near a window, I cocked my head, glanced at her, and sighed, wonderingly. She must feel like a vehicle, sent out to test-drive the law, I mused.

If the country and its healthcare system would get their day in court, if they were going to create clarity on the rules for dealing with medical error, then this was not going to help Mara. The black sheep would be herded through one more splendid spectacle of public judgment, but it was no longer about the sheep, if it ever was. It was about the principle. And she was merely its embodiment. When it was all over, whatever the outcome, she would have been used up, her purpose to larger interests played out, expired. A mere piece of detritus mangled through a criminal justice system in its quest for new turf, disposed once the flag had been planted. She would be remembered only in faint echoes of *schadenfreude* (thank God it wasn't me) and a waning trail of half-hearted collegial compassion ("We're so sorry for you, Mara"). The disillusionment with her work setting, her colleagues, her union, the justice system—the world—was etched on her face. Vindication would remain elusive. The truth would not come out.

But is there truth in the aftermath of a little girl's medication death? Or are there only versions?

At the Supreme Court

A few weeks after the meeting with the lawyer, I saw Mara again, this time in the ornate halls of the Supreme Court. High ceilings soared up, away from two tables, one for the defense and one for the prosecution. They were arranged in front of a regal podium decked out with a row of seats. When the justices had filed in and sat down facing both teams, the prosecutor reached for his version of the truth. I remember his craftiness, his cultural conformity to the conflict-avoidance of my adopted country. He was sitting down, not standing up. He was reading from a prepared script, not ad-libbing or grandstanding in front of his audience. His tone was measured, quiet, reverential. This is, I suppose, what court proceedings are supposed to do: separate emotion from substance, sublimate conflict into negotiation, turn revenge into ritual.

Mara sat over at the other table, flanked by her lawyer, with hands in her lap, eyes cast downwards. As she sat there, the prosecutor's opening statement started rolling over her, his gentle voice reverberating around the hall unamplified.

Except it wasn't a statement. It was a story.

"The baby was born on the 24th of February in the regional hospital," he intoned. He recalled the happiness of the child's parents and mentioned details to paint a picture of family bliss, soon to be disrupted by a treatment gone awry. "She had a normal birthweight, but showed some signs of seizures in her arm after delivery. Three days later, the seizures had become worse. She was given Fenemal, a cramp reducer. After stabilizing on the 5th of March, she was discharged. But less than a month later, the seizures came back. The infant was rushed to the emergency room and taken in for observation. Her Fenemal dose got increased to 5 milligrams per milliliter and she was discharged again two days later. The day after, her mother called the hospital. After consultation, the baby's Fenemal dose was increased again—over the phone—to a twice daily 2-milliliter portion of the 5 milligram per milliliter mixture. On the 22nd of April the baby was brought in as part of a routine checkup. Everything was normal."

He paused.

To recount his version of the truth, the prosecutor had created a narrative. Narratives are strong. He must have picked that up in class once, or in one of his many street fights. Or perhaps a story, or liking a story, understanding a story, is simply what makes us all human. Mara must have heard versions of the story hundreds of times now, I thought. She must have turned it over and over in her mind infinitely, picking away at her role, plaguing herself by retrospectively finding opportunities to not make a mistake, to not become the centerpiece of this imbroglio.

Act One was over. The justices looked at the prosecutor, silently. Spellbound or bored silly? It was difficult to tell. Time to set the stage for a plot twist. Act Two. A different ward: the intensive care unit (ICU). A new medication. And, of course, the introduction of the villain: the nurse.

"On the 12th of May, the baby was admitted with a new bout of seizures, and sent to the ICU. Her Fenemal was increased to 2.5 milliliters twice daily, and she even received a bolus dose of Fenemal. But the seizures continued. The baby was then given Xylocard, a lidocaine-based medication, at 2 milliliters per milligram. The seizures subsided. She was discharged again on the 16th of May, off Xylocard, and back on the previous dose of Fenemal. But on the 18th of May, her mother called the hospital to say that her baby was suffering a new onset of seizures, now lasting about five minutes each. In the evening, the child was taken to the hospital by ambulance and admitted to the pediatric ward. New seizures made that she was transferred to the ICU later that evening."

With the baby back on the scene of the crime-to-come, everything was ready for Mara to make her entry. The lines of the two lead roles could now converge.

"Early in the morning of Sunday, 19th of May, Mara showed up for work. There were not many patients in the ICU; things were quiet. The baby was doing better now. In preparation for her transfer back to the pediatric ward, Mara went to the medication room to mix the Xylocard solution."

He paused and picked up the two little cartons in front of him on the table. Then he waved them around.

"There, in the cabinet, were two packages: one containing an injection dose of 20 mg/ml Xylocard, and one with a 200 mg/ml Xylocard solution intended for intravenous, or IV, drop. Misreading the packages, Mara took the 200 mg/ml to prepare the baby's drop, instead of the 20 mg/ml, as was prescribed."

The chief justice motioned that she wished to see the packages. They were handed over. Passed from justice to justice, they were handled for what they were in that context: pieces of evidence in a manslaughter trial. The justices studied the packages with what looked like mild interest, but could just as well have been muffled puzzlement. What kind of evidence was this anyway? This was not just a common criminal instrument—a knife, a handgun, a fraudulent contract—these were pieces of highly specialized medication, excised from their normal surroundings of thousands of normal, similar-looking packages that form the backdrop of a nurse's daily life. Now these two boxes looked quite out of place, floating along the court's elevated regal bench, examined by people with little idea of what it all meant. Questions must have mounted, one on top of the other. What was it with these peculiar Greek neologisms, why were all these boxes white with green or light-blue lettering, and what were these befuddling volume–weight fusions?

The prosecutor continued. Not much longer now. Act Three. A rapid climax.

"That afternoon, back in the pediatric ward, the baby was hooked up to the new Xylocard drop, the one that Mara had mixed. But instead of subsiding, the infant's seizures quickly got worse. A pediatrician was called and tried to intervene. But nothing helped. Not long after, the baby was declared dead. Postmortem examination showed that she had died from lidocaine poisoning."

A story that makes sense, that is plausible, that has a powerful narrative arc and casts characters in recognizable roles of hero, victim, villain, and bystander can present a rather believable truth. And the prosecutor's story did. His plot painted a normal hospital, a normal, innocent little patient, attended to by normal physicians, suddenly all confronted by the sinister and totally unnecessary turn of events on a Sunday morning in May—the fatal denouement of Mara's mix-up. Quite impeccable. Quite logical.

A Calculation Gone Awry

But does that make it true? Consider another truth, the sort of "truth" that Mara had hoped in vain to bring out in the open on this day. After clocking in on the morning of May 19, she received a little briefing from the night ICU nurse. The original prescription had been unclear and not signed by the doctor who wrote it. The hospital (even the ICU) was equipped with a computerized prescription system, but the physician had been sitting at a terminal that happened to not be connected to the printer. Rather than moving to another terminal and print out a prescription, he wrote one by hand instead. Earlier that night the nurse had mixed a Xylocard solution with another physician's help, trying to divine the prescription. Now, in the morning, the doctor himself was asleep somewhere in the hospital, and, given that it was a quiet Sunday, nurses would not become popular by waking him up to ask a simple clarification. The night nurse showed the unsigned prescription and her medication log entry to Mara:

"40 ml + Xylocard 200 mg = 10 ml = 4 mg/ml, total of 50 ml"

"Remember, 10 milliliters Xylocard," the doctor had said to the night nurse, who now relayed this to Mara. The infant, in other words, had received a total of 200 milligrams of Xylocard by mixing two 100 mg/5 ml syringes (the standard injection package: 100 milligrams of Xylocard dissolved in 5 milliliters of fluid) into a 40-ml glucose solution. But in the ICU, syringes were never used for IV drops, because they contain a weak solution. Syringes were for direct injection only. The ICU used vials, with a stronger solution, for IV drops. But Pediatrics did not even have vials. They dealt with children, with little bodies that needed no strong solutions. Pediatrics routinely discharged the prepackaged syringes into an IV drop instead. The ICU seldom had little infants, though, and no tradition of using syringes for preparation of IV drops.

Later that day, when the night nurse had long gone home, Mara noticed that the infant's drop was running low and decided to prepare a new one. The baby would be transferred back to Pediatrics, but the move had gotten delayed. Remembering the "10 milliliters" reference from the doctor, and reading 200 mg off the medication log (as the prescription was unclear), she took two boxes that each contained a 5-ml vial with 200 mg/ml Xylocard. 10 milliliters total, and the figure of 200 mg—this was what the medication log said. She prepared the solution and wrote in the log

"Xylocard 200 mg/ml = 10 ml = 4 mg/ml"

Mara showed her calculations to another nurse and also to the Pediatrics personnel who came to collect the infant. The Pediatrics staff did raise a question, but it focused only on the dose of 4 mg/ml, not on the solution from which it supposedly would come. Five days earlier, when the infant had been in Pediatrics too, she had been on 2 mg/ml, not 4 mg/ml. The ICU confirmed to Pediatrics that 4 mg/ml was now the prescribed dose. The baby was to receive 10 milliliters of the solution that was supposed to contain 1 milligram of Xylocard for each milliliter.

But did it?

That night, Mara tossed in her bed. Her youngest son awoke a few times, rendering his mother restless. In the darkened bedroom, the events of the day came back to her. As far as she knew, the baby had lived; she had gone off shift before anything went awry. But something did not quite add up. Why had the night nurse, normally so assiduous, accepted such a messy and unsigned prescription? She had even had to call for help from a physician to mix the thing. And what about that log entry of hers? It had read "Xylocard 200 mg,",= but did that make sense? Xylocard 200 mg was meaningless by itself. 200 mg per what? Per…?

Mara sat up with a start.

Could it be true that she had taken two vials, instead of two syringes? They both contained 5 ml of fluid each, so any combination of two would amount to the 10 milliliters the doctor had wanted. The two packages were side by side in the cabinet which was so neatly organized on alphabet. But two vials meant…

She quickly ran the numbers in her head, peering into the darkness. Two 5-ml vials both containing 200 mg/ml Xylocard would have amounted to 2000 mg Xylocard, or 40 mg/ml, not 4! This would add up to a lot for a little infant. Too much maybe. In that case her medication log entry didn't make sense either. Take 10 milliliters with each milliliter containing 200 mg, and you would not get 4 mg/ml. You'd get an order of magnitude more. Ten times more. Forty.

Why had nobody caught it? She had had people double-check! Pediatrics had checked! Also, an entry about the solution would have had to be made on the IV drop before it went into the child—another double-check. What had happened? She would try to figure this out as soon as she was at work again.

On her next shift, Mara asked about the little girl. "She has died" was the answer. Her heart must have sunk. But determined to figure out what had gone wrong, and if she may have had any role in it, Mara went to the binder with prescriptions and flipped back to Saturday night. Where was it? Where was the prescription, that messy, unsigned prescription, that her predecessor night-nurse had interpreted as "200 mg" Xylocard, setting her, Mara, up for a possible mistake?

The prescription was gone. It wasn't there. It had disappeared and would never be found again.

Years later, only a few weeks before the hearing at the Supreme Court, Mara would plead with her lawyer to bring up the missing prescription. He yielded not an inch.

"How can you bring up something that doesn't exist?" he asked.

"But," Mara countered, "we are not allowed to prepare medications without a prescription, there *has* to be a prescription, and in this case there was too. Somebody took it out!"

The lawyer sighed and was silent.

"Look," he said after a while. "This is not the time to introduce new evidence. And even if it was, you can't produce as evidence something that you don't have. It's that simple."

Mara's world must have spun around her. She was locked up inside a Kafkaesque entanglement that had erased any resemblance to the real world. Her mind must have cast around for anything stable, anything recognizable, anything sensible. Instead it was finding nothing to grab onto, no lifeline, no help. And no "truth."

"Mea Culpa"

What there was, and what had been introduced as evidence, of course, was her own medication log entry. The one that said that 10 milliliters of fluid, with each milliliter containing 200 milligrams of stuff, would amount to a measly 4 milligrams of the stuff per milliliter in a 50-milliliter IV drop. It wouldn't. It would yield ten times as much. She had recorded her own miscalculation—putting the truth out there, for all to see.

Not long after learning of the baby's death, complying with reporting procedures in the hospital, she wrote to her superior:

> When I was going to mix Xylocard at around 10:45 that morning, I looked at the prescription and got Xylocard 20 mg/ml. I read both the package and the vial and recall that it said 20 mg/ml. I looked at what was prescribed and what I should prepare. So I got 20 mg/ml which I mixed with glucose 5% 40 ml.
>
> I asked another nurse to double-check but did not show her the empty vials. Then Pediatrics came to get the infant, ... and they took my prepared solution with them to hook it up in their ward. When the infant left us at 11:07, there was still about 3 ml in the previous drop, which had run through the night of the 18–19th of May.
>
> The following night, I awoke and suddenly realized that a vial normally contains 1000 mg/5 ml. And I had thought that I drew a solution of 20 mg/ml. When I was working the following Wednesday, I got to hear that the infant had died. I then understood that it could have been my mistake in making the solution, as there are no vials of 20 mg/ml.

Stories of mistake can be so simple. "My mistake," Mara had recorded. Mea culpa. To many others in the hospital, such an unprovoked admission must have been a godsend. Not that they would ever say, of course. They would not have to. The legal aftermath itself would prove them right. Mara was in the dock. Again and again. Nobody else.

Not that this would necessarily feel natural to anyone involved in the saga as it unfolded. Take a story as experienced from another nurse's point of view. When the infant started to show an increase in seizures and other problems after being hooked up to Mara's IV preparation in Pediatrics, nurses called the attending physician.

He responded by phone: "Up the flow, give her more."

They did. The problems got worse.

They called again. "Give her more, give her a bolus dose" was the instruction again. They did.

But this did not seem to help at all—in fact, things were going from bad to worse very quickly now. The attending anesthetist was now called by phone, but nobody answered. Another was found by calling through the intercom, but nobody showed. Only minutes later did the attending pediatrician show up in person. He ordered another bolus dose of Xylocard, but this had no effect either. The baby now needed 100% oxygen but she started vomiting into the mask, exacerbating her respiratory problems. The pediatrician ordered another bolus dose of Xylocard, thinking that this would finally stop the seizures. Then, during one attack, the girl presented with respiratory failure. The pediatrician responded by intubating the baby, and cleaned the airways by suction. Then the anesthetist arrived. The baby was ventilated but the suction tube proved too narrow for her passages to be cleared. Another bolus dose of Xylocard got pumped into the IV port. Finally, a thicker tube was found and inserted, clearing her airway. It was all too late. The infant went into circulatory shock. Adrenaline, Atropine, and Tribonat were given; heart massage administered; even the defibrillator was pulled out. To no avail. The baby was declared dead not long after. A postmortem showed that the girl had ended up with 43 micrograms of lidocaine per gram of her blood. The therapeutic dose is less than 6 micrograms per gram of blood.

Even if Mara had mixed from the 20 mg/ml syringes and not the 200 mg/ml vials, the infant would still have ended up with twice the therapeutic dose due to the volley of bolus shots during her final moments. Yet that is but one "truth" too. See the world from the pediatrician's perspective and another sensible story swims into view. The initial symptoms of lidocaine poisoning can include (yes) seizures. So the symptoms of too much Xylocard and too little Xylocard would have been similar, setting the physician onto a compelling plan to continue. Strong initial cues suggested his response was the right one. They had been confirmed before: this baby had responded well to the treatment of her seizures with lidocaine. The dose had been upped before, with good therapeutic consequences. He knew all this. And, for that matter, he never knew that the IV drop was administering the drug at 10 times the ordered rate. The quality of his assessments and decisions could impossibly be rated against that knowledge—knowledge he did not possess at the time. That much would be "true."

But what did the doctors actually know? I remember Mara countering this even before the final trial. Did they ever diagnose the source of the spams? Mara would ask. No, they didn't. Did they have any idea why the child responded better to Xylocard than to Fenemal, even if Xylocard is not mainly intended to deal with seizures?

Did anybody ever think to call in a neurologist? No. Did they ever ask themselves why the baby would suddenly develop such intense symptoms after getting back to Pediatrics on Sunday afternoon? Not that Mara knew. Did they ever recognize their own role in the slippage of prescription routines? In taking a nap at work on a quiet Sunday morning and being really grumpy when awoken for no apparent good reason? In not bothering to get up and mosey 10 feet to another computer to print out a prescription for Xylocard, rather settling for a bunch of handwritten squiggles instead? In not showing up for many, many critical minutes when a little baby was suffocating in her own vomit, wasting away on some drip? And then giving order after order after order of poisoning lidocaine? No, not that Mara would be aware. And who took that prescription away after the baby died? Where was it? And whose idea was it to start swapping a baby between Pediatrics and the ICU, a ward designed in every way for taking care of big people, not little ones? Were any of those "truths" ever going to be brought out?

CRIMINAL LAW AND ACCIDENTAL DEATH

A legal system holds people accountable. But it does not allow people to hold their account. Mara had become a hostage of legal procedure and protocol, and she decried the shackles on what she was granted to say and when. At every turn in the legal plot, she went in to battle the limits, to break through the constraints. She wanted permission to give her account. She just wanted the "truth" to come out. But at every end, she came out broken herself. Her account would still be inside of her—biting, festering. And increasingly bitter and partisan.

A legal system constructs an account from its own pick of the evidence. It makes its own story. It is interesting that society may turn increasingly to their legal systems to hand out that story, to provide accountability after a terrible outcome. There must be something in that account that we find terribly attractive; more enticing than what the people have to say who were actually there. Mara, for example.

Of course, we could dismiss their accounts as exculpatory, as subjective, biased, ulterior. Still struggling to understand her own performance, Mara had told a lower court that she may have misread the package labeling. By the time she got to the Supreme Court, however, she indicated that this was probably not the case: she mistakenly believed that 200 mg/ml was what she needed to have. This would certainly have made sense, given the prominence of the figure 200 in the medication log, and the reminder to end up with a volume of 10 ml Xylocard in total. But look at how the Supreme Court chose to interpret the various accounts that Mara had tried to provide. Put up as a last-grasp attempt to exonerate herself, to "find an explanation afterward," the Supreme Court painted Mara as ditzy when it came to getting an account of what had happened that Sunday in May:

> During the court proceedings, the ICU nurse described multiple ways how it could be that she mixed the IV drop with the wrong concentration of Xylocard. What she offered cannot therefore express what she really remembers. Rather, her accounts can be seen as attempts to find an explanation afterward. They are almost hypothetical and provide no certain conclusion as to why she did what she did.[69]

Whatever Mara offered, the sheer variety of her accounts had disqualified her as a purveyor of truth. In her stead, the Supreme Court was happy to provide the "certain conclusion" so sorely lacking from Mara's story. They speculated why Mara did what she did: she "misread, miscalculated, or took the wrong package" from the shelf—all because of "negligence." Mara did what she did (whatever it was), because she was careless. "She could have read the medication log more carefully, calculated more carefully or done any other double-check that would have revealed her error and its potentially fateful consequences."[69] But she did not. She was negligent. In the absence of a story from Mara that made sense, people turned to the legal system to serve them a story with a cause and a culprit. The cause was misreading, miscalculating, or grasping wrong due to negligence, and the culprit was Mara. Instead of listening to the protagonist, people legitimated a particular institution to get at the "truth" and mete out supposedly appropriate consequences. They may have thought, as many increasingly do, that this legitimated authority could deliver the veridical account—what really happened. For the one who was there could not be trusted to deliver an account that "expressed what she really remembered." She, after all, could "provide no certain conclusion as to why she did what she did."

Of course, judicial proceedings do rely on the insider account as part of their evidence base. Mara was given a voice—here and there. But she never called the shots. She spoke when spoken to: merely proffering a hunch of answers to often inane questions gurgling from a tightly scripted ritual:

"So what did you read on this package, did you read anything at all, or did you take fluid directly from the vial?" the prosecutor in the Higher Court had insisted.

"I looked at both the package and the vial," Mara had replied.

"What did you see?"

"I don't know. I wrote 200 mg per ml, but I don't know."

"You don't know."

"No."

It sounded exasperated—feigned or real: "You don't know." If Mara did not know, then who would? She had been there, after all. Again, the inability to give that final account, that deeper insight into the workings of her own mind that day, was taken as reticence, as foot-dragging. "You don't know" was taken, as it often is by the time adversarial positions are lined up in a criminal trial, not as "you really don't know," but as "you don't want to tell us."

I recall sitting in the lawyer's office with Mara when she offered the explanation in which she really believed she had taken the right package, the one she was supposed to take (as that was always the one she prepared IV drops from). There was no misreading; that had been a wrong explanation. But the Supreme Court justices would have none of that. They would not see the latest account as a genuine attempt of the insider to articulate what had happened, but as a ditch from the debris, as a ducking of responsibility.

RATIONAL SYSTEMS THAT PRODUCE IRRATIONAL OUTCOMES

And so we turn to our legal system to furnish us with the truth. Deploying rational techniques like those of a trial, rather than institutional authority (such as that of the Church) putatively allows us to arrive at true accounts and appropriate moral rules. But intense attempts at deploying rationality, sociologist Max Weber warned over a century ago, quickly deliver the opposite. The output of supposedly rational institutions is often—quite naturally, necessarily—irrational. There were many, both inside and outside the healthcare system, who thought just that about Mara's verdict. When a nurse herself reported a mistake, in an honest effort to abide by the rules and perhaps help prevent recurrence, it made no sense at all to have her end up convicted of manslaughter for the very mistake she voluntarily divulged. This was irrational. Even more poignantly, why she? Singling out Mara for this adverse outcome of a discontinuous, wandering processes of care delivery that counted many contributions from many contributors made no sense whatsoever. And then, this was not the first or only medication adverse event ever, not a uniquely egregious occurrence. In the same year that Mara was first charged, more than 300 severe medication errors were reported to the country's health authority. Adverse medication events are "normal." They are the rule, or at least part of it, baked into the very fabric of delivering assorted compositions of volumes and weights and rates of substances through various means. This, moreover, is accomplished through a thoroughly discontinuous process, where gaps in the delivery of healthcare open up because of changes of medium (e.g., from oral to written to oral prescriptions or dosage orders), handovers from one caregiver to another between shifts, movement of patients between wards, transferal of the caretaking physician, or other interruptions in workflow. Patients, prescriptions, orders, medications, and healthcare workers all cross departments, shift responsibilities, flow through hierarchies, and traverse levels of care as a matter of routine. It would be easy, then, and quite rational, to show that Mara's adverse event was part of a systemic feature of healthcare delivery. So how a supposedly rational judicial process could come to the exact opposite conclusion is something that Weber would not have found surprising. The accounts of human error that a legal system produces can be so bizarre precisely because of its application of reason: the ways judicial proceedings rationalize the search for and consideration of evidence, closely script turn-taking in speech and form of expression, and limit what is "relevant" are institutionally constrained in their deferral to domain expertise, and necessarily exclude the notion of an "accident" because there is no such legal concept.

When you come up close, close enough to grasp how case content becomes subjugated by judicial form, close enough to hear the doubts of the victims about the wisdom of having a trial in the first place, close enough to taste the torment of the accused, to feel the clap of manacles around the expression of their own account, to experience the world from the dock and sense the unforgiving glare it attracts, a more disturbing reality becomes discernible. In the view from below, there is a deep helplessness: an account is created by nonexperts who select bits and pieces, in a process that runs its own course and over which there is very little—if any—external control. To those present when the controversial event happened, and who may now be in the dock (as well as to many of their co-practitioners), the resulting account may well be bizarre, irrational, absurd. And profoundly unfair.

THE SHORTEST STRAW

Mara had hoped that the process in the Supreme Court would end up bringing out a real version after all the acrimony in lower courts. It did not. Instead of truth, she got an upheld conviction. Instead of vindication, she got something that she could not possibly consider "true" anymore.

Sitting in the twilight of her living room on a rainy day late in August, months after the hearing, I began to believe that her psychological devastation was due not just to the Supreme Court upholding the guilty verdict, including its heavier penalty. This may not even have been the chief source of her anguish. With her license to practice still intact, and the sentence turned into conditional time, it had few overt practical consequences (not that she could, or wanted to practice in the ICU ever again, by the way). No, I started to sense rather a resignation, a disillusion, a dizzying realization that progress toward truth is not a movement from a less to a more objectively accurate description of the world. She might have hoped that we all could learn the truth behind the death of the little girl. But there is no such truth to find, to arrive at, to dig out. No final account, no last word—only versions, jostling for supremacy, media-light, popular appeal, legal sustainability. And her version had consistently drawn the shortest straw. Again and again.

4 The Criminalization of Human Error

Aviation and healthcare, as well as other fields of safety-critical practice, are reporting an increase in the criminalization of human error and criminal prosecution in the wake of an aviation accident. Assigning such liability is currently standard practice in many countries. Italy has a specific criminal category of causing "air disaster," and two airline pilots were recently sentenced to 10 years in prison after a crash that killed 19 people. In aviation, criminal prosecution of mostly front-line operators in the wake of incidents and accidents has occurred in the Netherlands, England, Spain, France, Italy, Greece, Cyprus, the United States, and Taiwan, as well as other countries. In healthcare, Sweden recently debated the introduction of the category "patient safety crime."[70]

The judge investigating a 2008 Madrid air crash that killed 154 people has called three mechanics for questioning on suspicion of manslaughter. The two technicians who checked the plane and cleared it for takeoff on August 20 and Spanair's head of maintenance at Barajas Airport are facing charges of 154 counts of negligent homicide for failing to detect faults that led to the tragedy.

Judge Javier Perez has launched a judicial investigation, independent from that of Spain's Civil Aviation authority, to determine the causes of the crash. Spanair flight JK5022 bound for the Canary Islands crashed on its second takeoff attempt after the wing flaps failed to deploy. An alarm system in the cockpit failed to warn pilots of the fault and the twin jet engine rose about 40 feet before it veered to the right and slammed into the ground tail first. The back of the aircraft broke apart and the fuselage bounced three times before crashing into a shallow ravine and bursting into flames. Only 18 people survived Spain's worst air accident in 25 years.

The preliminary report by Civil Aviation investigators absolved the pilots of any blame for the accident after data from the black box recorder showed they had followed the correct procedures. Judge Perez was investigating whether the maintenance crew charged with repairing an earlier fault that led to the first takeoff attempt to be aborted were negligent in making the necessary repairs. It appeared that they may not have checked whether a problem detected in an air temperature gauge on the outside of the aircraft was caused by a mechanical fault that affected other parts of the plane. It has also emerged that the same aircraft suffered problems with wing flap deployment on two occasions in the days leading up the crash.[71]

Concern with the criminalization of mistake exists in safety-critical domains beyond aviation and healthcare. This includes shipping, construction, and chemical processing. The laws under which criminal prosecution of professionals currently occurs are mostly derived by extending general hazard statutes from particularly road traffic laws that criminalize the reckless endangerment of other people or property. The move to criminalize human error (a label that is itself a psychological attribution) could parallel the evolution of, for example, laws on hate crimes, which went from a broad, ambiguous category to a focused, determinate legal construct.

Doubts have been raised about the fairness of criminalizing errors that are made in the course of executing normal professional duties with no criminal intent and the capriciousness of criminal prosecution. Remember Mara: as a nurse, she was criminally convicted for a medication administration error of a kind that was reported to the regulator by others more than 300 times that year alone. Doubts also exist about the ability of a judiciary to make sense of the messy details of practice in a safety-critical domain, let alone resist common biases of outcome knowledge and hindsight in adjudicating the professional work of practitioners.

Seven Tuninter pilots, technicians, and managers were convicted by an Italian tribunal for a crash in which 23 passengers survived and 16 were killed in August 2005, Andy Nativi wrote in Aviation News Release *(March 25, 2009). The crash involved an ATR-72 that took off from Bari, Italy, that was headed to Djerba, Tunisia, and was forced to ditch in the sea just along the Sicilian coast.*

The incident occurred because the aircraft fuel gauges and indicators had been incorrectly replaced by the maintenance personnel with those of the ATR-42. The instruments indicated there was enough fuel on board when the aircraft took off, although there was not actually enough fuel to carry out the intended flight.

The two pilots were charged with multiple counts of manslaughter and air disaster, and sentenced to a term of 10 years because, in theory, they had the opportunity to reach the Palermo Airport for an emergency landing if they had followed proper procedure. Instead, the pilots chose to pray. Another five technicians and managers were found guilty, with the chief operating officer and the maintenance chief sentenced to nine years each, while three technicians were sentenced to eight years each. Two others defendants were not found guilty. None of those indicted were present at the tribunal, and Tuninter lawyers announced they would appeal the verdict.

Let's first look at who are involved in the criminalization of human error.[72] Who are the parties that make it so? Whose interests are at stake, and where do they collide? Then let's consider some of the evidence of what criminalization does to safety and ultimately raise the question how a society can support a way of dealing with risk that perhaps works against its own long-term interests.

THE FIRST VICTIMS

Those who should benefit foremost from any legal action in the wake of an incident or accident are those who were affected most by it: the first victims. Take the parents of the infant in the case study just prior to this chapter. Or the family of a passenger in an airliner crash.

Most countries give victims the role of factual witness in a trial (in addition, a so-called witness-impact statement has become popular in the US, allowing witnesses to reveal to juries even the emotional and other toll taken by the accident or "crime." There are suggestions that this can unfairly sway juries against the suspect).

Witness testimony can highlight all kinds of angles on the case, from the emotional and practical consequences of the loss they suffered to their own observations of the behavior of, for example, doctors or other caregivers before, during, and after the occurrence. This can bring aspects to light that would not otherwise have been known, as victims may sometimes have had a close-up view of the unfolding incident.

This in turn can put judges (or juries) in a tricky position. If court cases are conducted on the presumption of innocence, and scrupulously avoid the word "perpetrator," instead using a more tentative "defendant" or "suspect," then what does this do to the status of the victim? And what does the victim do to the status of the defendant or suspect?

If you follow this logic, then without a proven crime, there can be no victim of a crime. A strong validation of a victim's account in court, then, could perhaps make people lose sight of the difference between suspect and guilty. It can become difficult to remain unbiased and retain the presumption of innocence.

But having victims' testimony in court serves other important functions that may sometimes outweigh these risks. Giving testimony in court offers victims an opportunity to get their voices heard. This is very important. They often want to tell their story, or parts of it at least. As explained in Chapter 2, if no other provisions are in place, then a lawsuit or a trial may be the first time that anybody bothers to listen seriously to the victim, and the first time the victim has the opportunity to hear the story told by those who were involved.

Such blocked yearning may be one of the grounds for going to court in the first place. Michael Rowe, a sociologist at Yale, captured this in an essay about his son's death after two failed liver transplants: "Many of those who sue doctors ... have no place else to hand their grief when that grief—and seemingly their loved one's life—is being ignored, even declared, in the space left by silence, a thing of no value.[73]

Yet not everybody turns to courts to get their voices heard. Victims can turn to the media instead. These can provide an outlet for what could end up a rather one-sided account of the incident or accident.

Remember from Chapter 2: What matters for an organization involved in a tragic incident is to validate first victims' concerns and wants, and to do this promptly. Not all organizations have well-developed response mechanisms in place that deal in a respectful and timely manner with the needs of victims. A basic desire of the victim is simply to be recognized, to get a chance to tell his or her side of the story. And to not have to wait for months or to force the organization to listen.

Do First Victims Believe That Justice Is Served by Putting Error on Trial?

If a trial is the first time for victims to tell their story, then that part of justice will likely be seen as served. But what about the consequences for the accused? Do victims see those as just? The record may be surprising. In nurse Mara's case, the mother of the infant began to doubt both the point and the fairness of the trials against one ICU nurse well before it was all over.

This is one reason why victims can have doubts about putting practitioners on trial for their alleged errors. The organizations that helped produce the problem are often left untouched; the norms, values, policies, and regulations that drive their business are not critically examined. Putting the front-end operator on trial is an example of single-loop learning, which focuses on the first part (possibly a human) that can be connected to the failure and replacing or otherwise dealing with just that part.

A judge on Monday convicted four people and acquitted three in a 2001 plane crash that killed 118 people at Milan's Linate Airport in Italy's worst civil aviation disaster, a lawyer for victims' relatives said. The four were convicted of multiple manslaughter charges and sentenced to prison terms of up to four years and four months, said the lawyer, Alessandro Giorgetti. Three of the defendants were employees of the Italian air traffic control agency and one was an airport official, Giorgetti said.

Early reaction from some victims' relatives in the Milan courtroom was lukewarm. "If we had to go by our feelings, no convictions would be enough," said Paolo Pettinaroli, who lost a son and is president of an association of victims' families. "But the law has decided this way and we accept it." He said, however, that the victims' families would appeal the acquittals.

It was the second verdict in the disaster. In a separate trial last year, an Italian court convicted four other defendants, including an air traffic controller and a former top aviation official, of multiple manslaughter and negligence and sentenced them to prison terms ranging from six and a half to eight years. The crash occurred on October 8, 2001, when a Scandinavian Airlines System airliner bound for Copenhagen and a corporate jet collided in morning fog on the Linate tarmac. The airport's ground radar was out of service at the time. The collision killed 110 people on the MD-87 jetliner, 4 people on the Cessna business jet, and 4 members of the ground crew. The ruling Monday came after a fast-track trial, which allowed a limited amount of evidence and provided for lesser sentences on conviction.[74]

For some victims a fast-track conviction of the practitioners involved can seem too easy, too quick, too convenient. It may even leave them on the sidelines: retributive justice typically does. It relegates first victims to mere bystanders, to an audience, not participants in the creation of justice. And such a conviction might also not get at the heart of the issue that animates many first victims: making sure that there is no next time. This is often one of the few recourses that victims have left. They have already been bereaved or injured by the incident or accident, and putting somebody in jail is not going to give them back what they lost. What uplifts instead is getting some confidence that it will not happen again, that somebody else will not have to go through what they had to suffer.

This confidence can perhaps evaporate when victims realize how a trial confines the remedy to a judgment about the right or wrong of only one person's actions. It does not get much better if the person in the dock is a manager or a director-general instead of an operator. Accused of deficient management or insufficient oversight, these individuals get to bear the full brunt of the diffuse failings of an entire system. Not many, not even victims, would see this as either reasonable or fair.

Are Victims in It for the Money?

What about financial compensation? Are victims interested in monetary compensation and is that why they will pursue or help pursue a trial? Not really. A recurring finding from lawsuits against physicians, for example, is not only that they are surprisingly rare, but also that patients or their families do not engage in legal action primarily because of money. They sue primarily to get the story out.[17] Patients and families do not typically engage in legal action until they have found that they are being stonewalled, that no "account" is forthcoming from the practitioners or organization involved in the adverse event. They want to hold the practitioners or organization involved accountable—literally, and initially often even without prejudice or desire for retribution. They want to hear the story from the side of the involved practitioners and their employing organizations: What went wrong? Where? And why? How can other patients or passengers or spouses of soldiers be protected from the same kinds of failures? These are often among the most pressing questions.

If there is no other route to such disclosure, people turn to the legal system as their final address for forcing out "accountability." Again, the "accounts" produced under such duress, of course, may have little to do with what happened and much more with protecting vested personal or organizational interests (see Chapter 3).

THE SECOND VICTIM

The person on trial (the second victim, but typically called a "suspect" in a criminal trial, and a "defendant" in a civil trial) really suffers two kinds of consequences:

- **Psychological**. The suspect or defendant may experience stigmatization and excessive stress, and feel humiliation, isolation, shame, and depression. Judicial proceedings occur essentially in a foreign language for practitioners of other professions, and they may feel very little control over what is going on or what the outcome may be.
- **Practical**. Practical consequences can include jail time or significant financial costs (fines, court costs, lawyers' fees). Such costs can be borne by insurance (in case of malpractice suits) and otherwise by professional associations (and sometimes by employers), because few practitioners (currently) have insurance that covers the cost of criminal prosecution. One other real consequence of criminal prosecution is the risk of losing the license to practice. A criminal or otherwise judicially tainted record is enough for some organizations to avoid a practitioner altogether. Loss of license often means loss of income, livelihood. It can mean loss of colleagues, context, familiarity, and perhaps loss of meaning itself. Some organizations that have the resources redeploy the practitioner, but not all have the wherewithal to do so.

Where licenses were not lost, the employing organization may still not dare to have the practitioner work operationally any longer, or the practitioner him- or herself elects not to. The nurse in the Xylocard case actually did not lose her license to practice as nurse. In a bizarre twist of legal protocol, the medical licensing board lost its access to and control over the investigation once the judiciary stepped in. They were never able to form their own judgment about the case or the nurse's ability to practice. She still has her license in her pocket today. But what does that mean? The nurse won't practice anymore. It is not likely that she could face a prescription from any physician that was even remotely unclear. It is unlikely that she could hook a patient up to any drop without asking herself a thousand questions. It is not likely that she could be effective anymore. Or safe.

In most countries, testimony of the suspect can be used as evidence in court. Interestingly, courts are mostly, or exclusively, interested in confessions. Denials are generally not seen as convincing. But if a suspect confesses the "crime," then this can be adequate for a conviction. No other evidence may be necessary.

What this means is that the police, or other investigating authorities, may sometimes have an interest in "helping" the suspect remember certain things, or state them in a certain way. Add to this that courts in some countries are content to review only a summary of the interrogation transcript, which may have been drawn up months after the actual encounter with the suspect, and the distance between what was intended and what can get interpreted by a judge or jury becomes huge.

It is not strange that, also for this reason, suspects may feel as if they are caught up in a Kafkaesque process. They get accused of things they do not know or understand, because these are cast in a language profoundly foreign from that which makes up their own world, their own expertise.

THE PROSECUTOR

Prosecutors are on the front line of defending and upholding the law. They have to decide which acts should be prosecuted. The role of a prosecutor is to launch a prosecution on behalf of the state.

WHAT TO PROSECUTE?

In the wake of an incident, whether to prosecute or not is often a very difficult call to make. In making this call, prosecutors could benefit from some guidance and perhaps even domain expertise. But access to objective domain expertise can be very hard. Whether to go ahead with prosecution or not is mostly at the prosecutor's discretion—in principle. In practice, there can be pressure from various directions:

- There may be political pressure to prosecute. Where prosecutors are elected, their constituencies could demand that they go ahead with prosecution. Where they are appointed, politicians could make clear in various ways that prosecution is desired (because politicians want to be seen as "doing something" about the problem).

- The role of the media can be significant here too: it could be that when the media calls for holding people accountable, then politicians may too.
- There can also be political pressure in the other direction (i.e., to not prosecute): some organizations and professional associations have lobbied successfully for agreements between politicians and other stakeholders, so that prosecutors leave professional incidents in particular industries alone.

Prosecutors in the Netherlands requested the flight-data recorder and cockpit-voice transcripts from the crashed Turkish Airlines Boeing 737-800 at Amsterdam Schiphol, David Kaminski-Morrow of Air Transport Intelligence News *wrote on February 28, 2009. The Openbaar Ministerie said the Dutch Safety Board does not have to grant the request, contained in a letter from its aviation department, because the prosecutor is "lacking" a legal basis.*

"We are awaiting a response from the Safety Board," said the prosecutor's office, adding that it submitted the request because the data can give "an insight into the circumstances of the accident." Judicial intrusion into accident investigation has attracted high-profile criticism, also after the Italian Agenzia Nationale per la Sicurezza del Volo (ANSV) accused legal officials of hampering the probe into a Cessna executive jet crash in Rome. ANSV called for deconflicting of investigators' needs from the rules of criminal proceedings. The US Flight Safety Foundation (FSF) also echoed ANSV's concerns, claiming prosecutors also interfered with French inquiries into a fatal Airbus A320 crash in the Mediterranean Sea. "In recent days the French authorities have returned some of the Airbus evidence to safety investigators," it stated. The FSF understands the demand for justice and accountability but added: "We cannot allow the safety of the aviation system to be jeopardised by prosecutorial overreach. Unless there is evidence of sabotage, law enforcement and judicial authorities need to step aside, allow accident investigators immediate access to the wreckage and to surviving crew and passengers, and let safety professionals do their job."

What to prosecute is clear—in principle. Just look in the law or jurisprudence. Yet in practice, and particularly in cases of "human error," it appears more random and unsystematic. One important reason is the sorts of laws used for such prosecution. Most stem from what could be called general risk statutes, which proscribe, for example, "endangering the lives" of other people. In many countries such statutes have their roots in road traffic law or laws governing damage to third parties in the normal course of daily life. Such laws are deliberately vague, and their jurisprudence predictably diverse, because of the infinite variation of situations that judges or juries may have to handle. But consider what happens when such general notions of danger or risk slide into considerations of culpability of practitioners' performance in a high-risk, safety-critical profession. Their very jobs involve the endangerment of the lives of other people.

Although safety data in many countries is unprotected because of freedom-of-information acts, prosecutors normally do not look into an organization's database in the hope of finding evidence of prosecutable acts. Something else must often rouse interest. A prosecutor may get a cue about the presumed seriousness of an error from the media, for example. This can be entirely coincidental, as in the Xylocard case, where the prosecutor stumbled upon her story in the local newspaper. Errors can sometimes be portrayed in the media as sufficiently culpable (even before any investigation) so as to capture a prosecutor's imagination.

Safety Investigations that Sound Like Prosecutors

Prosecutors and judges are not supposed to use official investigation reports in their judicial proceedings—at least this is the rule in many countries. There, the official investigation report cannot be used as evidence in court. But there is nothing in those laws that forbids prosecutors or judges from reading publicly available reports, just like any other citizen.

Over the past few years, I have talked to various investigation boards or departments about the language they use to describe people's actions in an incident or accident. All of these cases had attracted judicial interest. The people involved knew that judges and prosecutors were waiting for the formal report to come out (even though they were not supposed to use it formally in their judicial work).

If a trend toward criminalization is indeed happening, then recent safety board conclusions such as the ones about an aircraft accident that happened to two pilots on a repositioning flight, could be counterproductive. Unforeseen effects of high-altitude flying, for which the crew was not trained, made that they entered a stall and suffered a dual engine failure as well as other unfamiliar problems in their attempts to restart the engines (which had a history of in-flight restart problems). Cockpit procedures did not contain specific guidance on how to recover from the situation they had gotten into.

The transportation board, however, thought that "the pilots' unprofessional operation of the flight was intentional and causal to this accident...the pilots' actions led directly to the upset and their improper reaction to the resulting in-flight emergency exacerbated the situation to the point that they were unable to recover the airplane...the probable causes were the pilots' deviation from standard operating procedures, and poor airmanship."[75] Although such responses can be understandable (and may even be seen as justified), they are a little difficult to reconcile with the typical mandate of a safety investigation (which is not to find people to blame but to help prevent recurrence). Also, a focus on people's putative lack of professionalism and a direct link between their actions and the bad outcome can overshadow the more diffuse contributions, from inadequacies in training to general unfamiliarity with high-altitude operations, a history of engine restart problems, incomplete flight manuals, and a host of deeper organizational issues.

Perhaps language in investigation reports should be oriented toward explaining why it made sense for people to do what they did, rather than judging them for what they allegedly did wrong before a bad outcome. An investigation board should not do the job of a prosecutor.

THE PROSECUTOR AS TRUTH-FINDER

Countries whose laws stem from the Napoleonic tradition (sometimes called inquisition law) typically offer their prosecutors or investigating magistrates the role of "truth-finder." This means that they and their offices are tasked with finding all facts about the case, including those that acquit the suspect or mitigate his or her contribution. Just like a judge or jury, they have to presume that the suspect is innocent until the opposite has been proven.

Combining a prosecutorial and (neutral) investigative role in this way can be difficult: a magistrate or prosecutor may be inclined to highlight certain facts over others. Accusatory law (that stems from common law tradition), in contrast, actually assumes that a prosecutor is partisan. As shown in the case studies in this book, however, putting two versions of the "truth" opposite each other in an adversarial setting may still not be the best way to get to a meaningful, let alone honest, story of what happened and what to do about it. Also, the resources available to the two opposing parties may be quite asymmetric, with the prosecutor often in a better position.

Prosecutors can actually get access to evidence collected in safety investigations quite easily and use it in criminal cases. In the United States, National Transportation Safety Board (NTSB) investigators can be called to testify in civil cases, but only on factual information. They cannot be forced to offer their analysis or opinions about information collected in an accident investigation. There is no such restriction, however, when they are called to testify in a criminal court case.

There is also no legal restriction in the United States against the use of the actual tape from a cockpit voice recorder (CVR) in a criminal trial. This despite the Board's own extensive limits on CVR usage: it does not extend to other agencies. The NTSB, for example, strictly limits who is permitted to listen to the actual CVR tape, and these people cannot make notes of its contents. The NTSB does not release the recording or any copy of it and only makes public a transcript of the recording that is limited to details pertinent to the safety investigation. But those restrictions end at the NTSB's doors. There is no prohibition against criminal prosecutors issuing a subpoena for the CVR tape and using it in court.[76]

THE DEFENSE LAWYER

The defense lawyer has an important role in laying out the defense strategy of the suspect. He or she can, for instance, recommend that the suspect not answer certain questions, or not testify at all. Judges or juries are not supposed to draw conclusions about suspects' culpability if they choose to remain silent. But, consistent with the fundamental nature of social relations and accountability, such silence can get interpreted as a desire to skirt responsibility.

A real and practical problem faced by most defense lawyers is that they are unlikely to understand the subtleties of practicing a particular safety-critical profession. Nor may they really have to understand. Contesting that a particular action is culpable or not is grounded in legal interpretation, rather than a deeper understanding of the meanings of risk, normative boundaries, and acceptable performance as the insider would have seen them in that operational world at the time. Indeed, the legal terms that get people in trouble in court (such as "negligence") are not human performance terms. These things are worlds apart.

Defense lawyers can also be limited—in budget, in human resources, and in their authorizations to investigate—to dig up their own facts about the case. In contrast, prosecutors can for example deploy the police to force facts into the open (though even there, prosecutors often face competition for limited resources: others may want or need to deploy the police elsewhere). Prosecutors can sometimes draw on the resources of government crime labs, witnesses or forensic institutes. Defense lawyers instead often have to rely on voluntary disclosure of facts by parties that think it is their duty, or in their interest to help the suspect (the employing organization often does not, by the way). This is another reason why cases can get argued on legal rather than substantive grounds. Finding minor procedural or formal flaws that scuttle the prosecution's case can be a cheaper and more effective defense than trying to match the investment in lining up facts that prosecutors can usually make.

THE JUDGE

A judge in Napoleonic law generally has three tasks:

- Establishing the facts.
- Determining whether the facts imply that laws were broken.
- If laws were broken, decide adequate retribution or other consequences.

ESTABLISHING THE "FACTS"

The first task, establishing the "facts," is a really hard one. "Facts," after all, get assembled and then brought to the bench by different parties, foremost the prosecutor. Here the border between facts on the one hand and interpretations or values on the other begins to blur. Of course facts are disputed during a court case; this is the whole point of having a trial. But what a fact means in the world from which it came (e.g., a rule "violation") can easily get lost. Neither judges, nor many of the other participants in a trial, necessarily possess the expertise to understand the language and practice from a particular domain such as nursing or air traffic control. They do not know how that world looks from the inside, and were they given a chance for such a look, they may still not really understand what they saw (the legal teams in the air traffic control case from Chapter 3 were given such an opportunity, and the nurse's judges were given the Xylocard packages to look at). What the facts meant in context can remain hazy.

For this reason, judges sometimes rely on outside experts to help them decode the facts that are delivered to the bench. This is where expert witnesses come in: other practitioners or perhaps scientists whose field is relevant to the issue at hand. But judges and prosecutors and lawyers often want to ask questions that lie outside the actual expertise of the witness. Either the expert witness must decline answering, or indicate that she or he is not really confident about the answer. Neither is likely to bolster their credibility or usefulness in a courtroom. Expert witnesses are supposed to be friends of the court, that is, help the judge understand the facts from an unbiased point of view. But witnesses are selected by one of the parties, and neither party is obliged to disclose how long they looked around to find an expert witness whose opinion was favorable to their side of the story.

Determining Whether Laws Were Broken

Determining whether the facts imply that laws were broken is at least as difficult as establishing the facts. How does a judge move from the facts to this judgment?

Scientists are required to leave a detailed trace that shows how their facts produced or supported particular conclusions. Such a trace typically involves multiple stages of analysis. The researcher shows, for instance, how he or she moved from the context-specific empirical encounter (the "facts") to a concept-dependent conclusion. What scientists know, in other words, cannot be taken on faith: they have to show how they got to know what they know. This is hammered into the rules of the game; it is part of the prerequisites for publication.

For judges, however, such burden of proof does not seem to exist to the same extent. How they believe that the facts motivate a particular conclusion (and thereby judgment) can be expressed in a few lines of text.

Is a jury any better at this than a judge? Law based on Napoleonic principles does not use a jury to move from fact to judgment (nor to decide on punishment), but common law typically does. A jury takes away few of the problems that the judge faces (they are not likely trained in the practitioner's domain either; establishing facts and basing a judgment of unlawfulness on them is probably difficult for them too), and also introduces new problems. One is the peculiarities of group behavior, from groupthink to the emergence of a dominant jury member. Jury selection is another, especially where jury members get selected on how they will likely vote on particular aspects of the case. And the resulting group is unlikely to be a "jury of peers" where the peer to be judged is somebody who exercised a complex safety-critical profession that required many years of specialist education and training.

Deciding Adequate Punishment

Professionals convicted of wrongdoing often do not end up in jail, or not for a long time. Judges do seem to conclude that this is not going to be rehabilitative. Fines or conditional sentences may be given instead. Of course, neither is likely to help improve safety in the domain from which the practitioner came, and they may not even be seen as just either.

The former captain of the Costa Concordia cruise liner was sentenced to 16 years in prison on Wednesday for his role in the 2012 shipwreck, which killed 32 people off the Tuscan holiday island of Giglio. Francesco Schettino was commanding the vessel, a floating hotel as long as three football pitches, when it hit rocks off the island, tearing a hole in its side. A court in the town of Grosseto found him guilty of multiple manslaughter, causing a shipwreck, and abandoning his passengers in one of the highest profile shipping disasters in recent years.

However, the judges rejected a request that Schettino begin his sentence immediately. They ruled instead that would not go to prison until the appeals process is completed, which can take years. The captain wept during his final testimony on Wednesday but did not return to the court to hear the verdict. Prosecutors had asked for a prison sentence of 26 years for Schettino, who has admitted some responsibility but denied blame for the deaths that occurred during the evacuation. Some lawyers representing the victims said the sentence was inadequate.

Investigators severely criticized Schettino's handling of the disaster, accusing him of bringing the 290 meter-long (950 feet) vessel too close to shore. The subsequent shipwreck set off a chaotic night-time evacuation of more than 4000 passengers and crew. Ann Decre, the head of a body representing French survivors, said the verdict could not cover the human cost. "For me it's six months for each death. And the family of the dead people, it's not six months or seventeen years for them, it's forever," Decre said outside the theatre that was turned into a makeshift courtroom.

Schettino was also accused of delaying evacuation and abandoning ship before all the 4229 passengers and crew had been rescued. He said earlier in the trial that he had been thrown off the ship as it tilted. The court sentenced Schettino to 10 years for multiple manslaughter, five years for causing the shipwreck, and one year for abandoning his passengers. He also received a one-month civil penalty for failure to report the accident correctly. He was left alone on the stand to answer for the disaster after the ship's owner, the Costa Cruises unit of Carnival Corp, paid a 1 million euro ($1.3 million at the time) fine and prosecutors accepted plea bargains from five officials. "Lots of people who were there and played a role were excluded today," Schettino's lawyer Donato Laino said outside the theatre. "We think the facts of the case were different." He and Costa Cruises were jointly ordered to pay a total of 30,000 euros each in compensation to many of the ship's passengers as well as millions of euros in compensation to Italian government ministries, the region of Tuscany and the island of Giglio for environmental damage. Earlier on Wednesday Schettino had rejected prosecution accusations he had shown no sense of responsibility or compassion for the victims, saying "grief should not be put on show to make a point."

The massive hulk of the Costa Concordia was left abandoned on its side for two-and-a-half years before it was towed away in the most expensive maritime wreck recovery in history. The last body was not recovered until 2014. Schettino's defense team argued he prevented an even worse disaster by steering the ship close to the island as it sank. Many other Costa masters also made "flyby's" of Giglio and other islands, as it was a commercially attractive thing to do. They said the sentence that was sought by prosecutors went beyond even sentences sought for mafia killers.[77]

LAWMAKERS

Lawmakers do not have a direct stake in legal proceedings or what it does to the creation of just cultures—other than the stakes they represent for their constituencies (voters). But legislators do play an important role, as they are eventually the ones who help sketch out the lines in laws that will then be drawn more clearly and applied by prosecutors and judges. They may also have a stake in aligning national laws with those of international bodies. Employing organizations or professional organizations may find that without some type of access to relevant legislators, making changes in the direction of a just culture could be difficult.

THE EMPLOYING ORGANIZATION

At first sight, employing organizations would not seem to benefit from the prosecution of one of their practitioners. It often generates bad press, the brand name can get tarnished, and management can be made to look bad or incompetent in the media.

On the other hand, employers can sometimes feel that they have to protect vested organizational interests, which may involve a degree of defensive posturing and shifting of blame.

What can get lost in the struggle to handle the immediate stress and challenges of legal proceedings is the organization's ethical mandate. This is, for example, to create safety (such as air traffic control) or to care for people (a hospital). Creating safety means not relying on simple, individual explanations for failure. Implicitly or explicitly supporting simplistic accounts of a bad apple could be seen as violating the very mandate the organization has. And why would that mandate not extend to the period after an accident that exposed the opposite? Caring for people means not discarding a nurse or doctor during or after he or she has been made to carry the blame for failure.

Some professions have come quite far with the development of so-called crisis intervention, peer support, or stress management programs that are intended to help practitioners in the aftermath of an incident. The importance of such programs cannot be overestimated: they help incidents become less of a stigma, that they can happen to everybody, and that they can help the organization get better if the aftermath is managed well.

THE CONSEQUENCES OF CRIMINALIZATION

MOST PROFESSIONALS DO NOT COME TO WORK TO COMMIT CRIMES

In considering the stakes of the various parties involved in the legal pursuit of justice, it is important to remember that most professionals do not come to work to commit a tort or a crime. They do not come to work to do a bad job at all. Their actions make sense given their pressures and goals at the time. Their actions are produced by and within a complex technological system, and are part and parcel of a normal workday. Professionals come to work to do a job, to do a good job. They do not have a motive to kill or cause damage. On the contrary: professionals' work in the domains that this book talks about focuses on the creation of care, of quality, of safety.

IS CRIMINALIZATION BAD FOR SAFETY?

The sheer threat of judicial involvement is enough to make people think twice about coming forward with information about an incident that they were involved in.[18] Just imagine how the colleagues of nurse Mara may have felt about this. The nurse, after all, stepped forward voluntarily with her view on the death of the infant. As long as there is fear that information provided in good faith can end up being used by a legal system, practitioners are not likely to engage in open reporting.

Many admit that they will file a report only when there is the chance that other parties will disclose the incident (e.g., an air traffic controller may think that a pilot will report a close call if he or she does not do it; a nurse may feel the same way with respect to a resident physician present during the same event, or vice versa), which would make the event known in any case.

This puts practitioners in a catch-22: either report facts and risk being prosecuted for them, or not report facts and risk being prosecuted for not reporting them (if they do end up coming out along a different route). Many seem to place their bet on the latter: rather not report and cross your fingers that nobody else will find out either.

Practitioners in many industries, the world over, are anxious of inappropriate involvement of judicial authorities in safety investigations that, according to them, have nothing to do with unlawful actions, misbehavior, gross negligence, or violations.

Italian investigators are criticizing judicial authorities for becoming involved in air accident inquiries at an early stage, accusing them of hampering the process, wrote David Kaminski-Morrow in Air Transport Intelligence News.[78]

The Agenzia Nazionale per la Sicurezza del Volo (ANSV) has expressed its exasperation in the wake of a Cessna 650 executive jet accident on February 7. The ANSV said the investigation had "come to a standstill" after the seizure of crucial material, including the flight recorders and relevant "vital" documentation, by court and judicial authorities. It added that the US National Transportation Safety Board had requested a copy of the data on the recorders but that the ANSV, as a result of the development, had been unable to provide the information, adding that the delay had a knock-on effect on safety.

The ANSV said it wanted regulators to take "decisive action" to avoid the "problem of conflict" between investigation requirements and the rules of criminal proceedings. Cockpit crew representatives were supporting the ANSV's stance, and backed calls for a change in Italian law to give technical investigation priority over judicial inquiries. "Under present legislation the judicial inquiry takes precedence," said the International Federation of Air Line Pilots' Associations. Absence of data from the flight-data and cockpit-voice recorders from the Cessna jet will "seriously impact the effectiveness and speed" of the investigation, it added.

Operational organizations, and even regulatory authorities (which fall under departments or ministries other than justice—e.g., transportation) are concerned that their safety efforts, such as encouraging incident reporting, are undermined.[18] But what, exactly, are people afraid of? Judicial involvement can consist of

- **The participation of law enforcement officials in investigations**. There are countries in the developed world in which the police is witness or even participant in accident investigations (in, e.g., road traffic or aviation). This can impede investigatory access to information sources, as pressures to protect oneself against criminal or civil liability can override any practitioner's willingness to cooperate in the accident probe.

- **Judicial authorities stopping an investigation altogether** or taking it over when evidence of criminal wrongdoing emerges. This often restricts further access to evidence for safety investigators.
- **Launching a criminal probe** independent of a safety investigation or its status. Accident investigation boards in many countries say that this severely hinders their efforts to find out what went wrong and what to do to prevent recurrence.[79]
- **Using a formal accident report in a court case**. Even though using such reports as evidence in court is proscribed through various arrangements, these routinely get overridden or circumvented. And, in any case, nobody can prevent a prosecutor or judge from reading a publicly available accident report.
- **Getting access to safety-related data** (e.g., internal incident reports) because of freedom-of-information legislation in that country, under which any citizen (including the judicial system) has quite unfettered access to many kinds of organizational data. This access is particularly acute in organizations that are government-owned (such as many air traffic control providers or hospitals).
- **Taking the results of a safety inspection** if these expose possibly criminal or otherwise liable acts. This does not have to take much: an inspection report listing "violations" (of regulations, which in turn are based in law) can be enough for a prosecutor to start converting those violations (which were discovered and discussed for the purpose of regulatory compliance and safety improvement) into prosecutable crimes.

The safety manager for one organization told me how the person involved in an incident flatly refused that the incident be used for recurrent training, precisely because of the perceived risk of prosecution or other consequences. Even assurances of complete anonymity and de-identification of incident data were not enough to sway the practitioner. Although understandable, this denied colleagues an opportunity to engage in a meaningful lesson from their own operation. Normal, structural processes of organizational learning are thus eviscerated, frustrated by the mere possibility of judicial proceedings against individual people.

In all of these ways, judicial involvement (or the threat of it) can engender a climate of fear and silence. In such a climate it can be difficult, if not impossible, to get access to information that may be critical to finding out what went wrong, or what to do to not have it happen again. Here is another example of what that can lead to.

A prosecutor responsible for aviation decided to launch what she termed a "test case."[80] The crew of a large passenger jet on takeoff had suddenly seen another aircraft, being towed by a truck, cross the runway in front of them. Immediately they aborted their takeoff and stopped before reaching the intersection. Nobody was hurt. The air traffic control organization, as well as the country's independent transportation safety board, both launched investigations and arrived at pretty much the same conclusions. After unclear radio transmissions with the tow truck driver, an assistant controller had passed her interpretation of the tow's position to the trainee controller responsible for the runway. The assistant controller did not have a screen that could show ground-radar pictures. The trainee controller did, and took the position of the tow at the edge of the runway to mean that the crossing had been completed. Buttons on a newly added panel in the tower for controlling lighted stop-bars at runway intersections proved ambiguous, but at the time all looked in order, and he cleared the other jet for takeoff. Meanwhile, the coach of the trainee controller was performing supervisor duties in the tower. The account, in other words, was straightforward in its complexity: mixing elements of interface design, production pressure, weather conditions, handovers, short-staffing, screen layouts, and communication and teamwork—among many other factors. This, the safety community knows, is what organizational incidents and accidents are made of. Many factors, all necessary and only jointly sufficient, are required to push a system over the edge of breakdown. And all of those factors are connected to normal people doing normal work in what seems a perfectly normal organization. These factors, then, are also the stuff of which recommendations for improvement are made. And they were, also in this case. The Air Traffic Control organization issued no fewer than 23 recommendations, all of them aimed at rectifying systemic arrangements in, for example, design, layout, staffing, coaching, communications, and handovers. The independent safety investigation board issued nine, quite similar, recommendations. This, as far as the community was (and is) concerned, is how the incident cycle was supposed to work. A free lesson, in which nobody got hurt, was milked for its maximum improvement potential. The people involved had felt free to disclose their accounts of what had happened and why. And they had felt empowered to help find ways to improve their system. Which they then did, for everybody's benefit.

But two years after the incident, the aviation prosecutor of the country decided to formally charge the coach/supervisor, the trainee, and the assistant controller with "the provision of air traffic control in a dangerous manner, or in a manner that could be dangerous, to persons or properties." (The country's law actually contains such provisions.) Each of the three controllers was offered a settlement: they could either pay a fine or face further prosecution. Had they paid the fine, the prosecutor would have won her "test" and the door for future prosecutions would have stood wide open. The controllers collectively balked. A first criminal court case was held a year and a half after the incident. The judge ruled that the assistant controller was not guilty, but that both the trainee and the coach/supervisor were. They were sentenced to a fine of about 450 US dollars or 20 days in jail. The trainee and the coach/supervisor decided to appeal the decision, and the prosecutor in turn appealed against the assistant controller's acquittal.

More than a year later, the case appeared before a higher court. As part of the proceedings, the judges, prosecutor, and their legal coterie were shown the airport's tower (the "scene of the crime"), to get a first-hand look at the place where safety-critical work was created. It was to no avail. The court found all three suspects guilty of their crime. It did not, however, impose a sentence. No fine, no jail time, no probation. After all, none of the suspects had criminal records, and indeed the air traffic control tower had had its share of design and organizational problems. The court had found legal wiggle room by treating the case as an infringement of the law, as opposed to an offense. It was as if they were proving themselves right and wrong at the same time. The court was wrong to bring and prosecute the case because there was no offense, but did not waste tax money after all because they managed to find an infringement. This was actually a no-brainer, as an infringement means "guilt in the sense that blame is supposed to be present and does not need to be proven." The only admissible defense against this is being devoid of all blame. This would work only if the air traffic controller was off-duty and therefore not in the tower to begin with. It was a celebration of perverse formalism (to use judge Thomas' words): a decorous nod to the prosecutor who had gone out to test the waters, and a measly but still unsettling warning to air traffic controllers and other professionals that they were not above the law. And it stopped all appeals: appealing an infringement is not possible as there is no conviction of an offense, and no punishment. The real punishment, however, had already been meted out. It was suffered by the safety efforts launched earlier by the air traffic control organization, particularly its incident reporting system. Over the two years that the legal proceedings dragged on, incident reports submitted by controllers dropped by 50%.

Many people, especially from the various professional communities, are duly concerned. The secretary-general of the worldwide association of air traffic control providers warned of "grave and undesirable consequences for safety" when judicial systems get involved.[81]

But Isn't There Anything Positive about Involving the Legal System?

Some in the legal community see the criminalization of error as a long-overdue judicial colonization of rogue areas of professional practice. It is, they say, a clamp-down on closed, self-serving, and mutually protective professional "brotherhoods" that somehow assert a special status and hold themselves to be above the law. Law is seen as authoritative, neutral, and fair, and it should reign equitably over everybody (hence Lady Justitia's blindfold): there should be no exception or discrimination either way.[82]

An increasingly vocal consumer movement, wanting greater control over safety in a variety of products and services, has been seen as sponsoring this view.[83] Pilots, doctors, air traffic controllers—already adequately compensated monetarily for the responsibility bestowed upon them—should be treated like everybody else. If they commit a culpable act, they should be held accountable for it. Exceptionalism is antidemocratic.

There is no evidence, however, that the original purposes of a judicial system (such as prevention, retribution, or rehabilitation—not to mention getting a "true" account of what happened or actually serving "justice") are furthered by criminalizing human error.

- The idea that a charged or convicted practitioner will serve as an example to scare others into behaving more prudently is probably misguided: instead, practitioners will become more careful only in not disclosing what they have done.
- The rehabilitative purpose of justice is not applicable either, as there is usually little or nothing to rehabilitate in a pilot or a nurse or air traffic controller who was basically just doing her or his job.
- Also, correctional systems are not equipped to rehabilitate the kind of professional behaviors (mixing medicines, clearing an aircraft for takeoff) for which people were convicted.

Not only is the criminalization of human error by justice systems a possible misuse of tax money—money that could be spent on better ways to improve safety—it can actually end up hurting the interests of the society that the justice system is supposed to serve. Indeed, other ways of preventing recurrence can be much more effective:

Alan Merry dryly remarked: "The addition of anti-hypoxic devices to anesthetic machines and the widespread adoption of pulse oximetry have been much more effective in reducing accidents in relation to the administration of adequate concentrations of oxygen to anesthetized patients than has the conviction for manslaughter of an anesthetist who omitted to give oxygen to a child in 1982."[83]

If you want a people in a system to account for their mistakes in ways that can help the system learn and improve, then charging and convicting a practitioner is unlikely to do that.

TORT LIABILITY

So far, I have basically talked about criminal legal proceedings (and will do so again in Chapter 5). This has a reason: there may be a trend toward criminalizing human error. So it is useful to assess whether or not that is a reasonable way to achieve the dual goals of a just culture: explanations of failure that satisfy calls for accountability and offer opportunities for change and progress on safety. So far, the evidence suggests that criminal law does not contribute to the achievement of these goals.

But another kind, called tort (or civil) liability, has been in use to deal with human error for quite a while, particularly in healthcare. Tort is a legal term that means a civil (as opposed to a criminal) wrong. To be liable under tort law, you do not have to have a formal contract with the other party, as it covers duties for all citizens under a particular jurisdiction (which is true of criminal law too of course). If a court concludes that an action is a crime, then the state can impose punishment (such as imprisonment or fines). If an action is a tort, however, the consequence is usually the payment of damages to the party injured or disadvantaged by the action. Tort law is applicable particularly in legal systems that stem from English common law, but even Napoleonic and other legal systems have ways of compensating victims through civil legal procedures. The technical variations are of course both subtle and many. Also, there can be overlap between crime and tort in some countries: the same action can be prosecuted as a crime (possibly resulting in the state imposing penalties) *and* as a civil tort (possibly resulting in damages to the victim).

Tort law too, has come under criticism for contributing neither to safety nor to justice when it comes to human error[1,60]:

- Tort law is a very irregular mechanism to compensate victims of error. According to one study, only one in seven patients who can be said to have been "negligently" harmed ever gain access to the malpractice system. Those who are older and poorer are disproportionately excluded from access.[59,84,85]
- Tort law also delivers compensation inefficiently. Administrative costs account for more than 50% of total system costs, and a successful plaintiff recoups only one dollar of every $2.50 spent in legal and processing costs.[85]
- Malpractice claims offer only the chance of financial compensation. They do not have as a goal to encourage corrective action or safety improvements; they do not help people get an apology or any other expression of regret or concern.
- Tort law includes practices such as pre-trial discovery and all kinds of rules that govern disclosure and the protection of information. And of course, a trial is in itself adversarial, lining up people against each other in competitive positions. The upshot is that tort law makes it *more* difficult to get facts out, rather than helping people find out what went wrong and what to do about it so it does not happen again.
- Also, the adversarial process is based on the idea that the presentation of relentless, one-sided arguments to an impartial judge or jury is the best way to get to the "truth." Chapters 2 and 3 did acknowledge that multiple stories are necessary if we want to learn anything of value about complex events, but that does not mean only two, necessarily opposing stories, where what is true in one is almost automatically false in the other. The ones who tell these stories are often not the ones who know them best (the physician or the patient), but rather their lawyers, who will have to abstract away from the details and cast things in a legal language that can get far removed from the actual meaning of people's actions and intentions at the time.
- As with criminal trials (which do not deter people from making mistakes but *do* deter people from talking about their mistakes), tort law promotes defensive practice rather than high-quality care.[86]

WITHOUT PROSECUTORS, THERE WOULD BE NO CRIME

We do not normally ask professionals themselves whether they believe that their behavior "crossed the line." But they were there, and perhaps they know more about their own intentions than we can ever hope to gather. Perhaps they are in a better position to say whether substance abuse played a role, or whether the procedures that they violated were workable or correct or available. And whether they knowingly violated them or not. Yet we don't rely on insiders to give us the truth. After all,

- We suspect that those people are too biased for that.
- We reckon they may try to put themselves in the most positive light possible.
- We will see their account as one-sided, distorted, skewed, partial—as a skirting of accountability rather than embracing it.

To get a truthful account of what happened, we do not typically listen to the people who were there, even if we do sometimes give them a voice (like we do in a trial, for example).

The View from Nowhere

So, again, the central question keeps coming up: who gets to decide instead? Is there a perspective that is not biased? A perspective that is impartial, neutral? We often turn to our legal systems for this. We expect a court to apply reason, and objectivity, and come up with the real story, with the truth. And then hand out consequences for those responsible for the outcome. From a distance, it may well come across this way. A disinterested party takes an evenhanded look at the case. The appropriate person gets to be held accountable. Appropriate consequences are meted out. Truth and justice are served.

The legal system certainly goes to great pains to appear as if it is impartial. Many of the trappings of the justice system are designed to impart an image of rationality, of consideration, of objectivity and impartiality.

- Think, for starters, of Lady Justitia's blindfold—the very profile of neutrality.
- The pace of judicial proceedings is measured, the tone solemn.
- The rules of proceedings are tight and tightly controlled.
- The uniforms and settings and language invoke a kind of otherworldliness, of not exactly belonging to the daily, messy hubbub of the real world out there.
- Even the buildings are often designed so as to be set apart from the rest of the world: just imagine your own local courthouse. It is probably separated from the sidewalk by gates, lawns, forecourts, high steps.
- The judges are often behind enormous doors, seated at a distance from other people, on podia, behind solid desks, under high ceilings.

Does this symbolism and imagery, this elevation and separation—meant to offer the assurance of rationality and impartiality—really give a court a better, more neutral view of the truth?

There Is No View from Nowhere

Telling the story from an objective angle is impossible, no matter how objective, disinterested, unbiased you may think you are. Or how neutral we make Lady Justitia look with her blindfold. Just ask yourself, if you were to take an objective look at the world, from where would you look? An objective view is a "view from nowhere."[87] And there is no view from nowhere, as there would be nobody to form the view.

So no view can be neutral, or objective. Because no view can be taken from nowhere. This means that all views somehow have values and interests and stakes wrapped into them. Of course, we can try to control the influences of those values and interests. And the legal system has great traditions and symbols and rituals to do just that. But in the end, nobody can discover or generate a value-free truth. Judges are stakeholders in the healthcare system too. They may be consumers of it, after all. And they have a larger role: helping maintain stability, and confidence in a society's institutions.

In nurse Mara's Xylocard case, the Supreme Court admitted that its agenda was in part to reassure any disquiet about the safety of the healthcare system: "Concern for patients' safety and their confidence in the healthcare system, demand that the nurse's actions be seen as so clumsy that they imply culpable negligence. She therefore cannot avoid being responsible for manslaughter."[69] The maintenance of "confidence in the healthcare system" demanded the construction of a version where one antihero could be singled out to receive the blame, to bear the explanatory and moral weight of the infant's death.

For a court to find an offense, and to call it criminal, is not the product of blind arbitration. It is not the clearest view on things from an objective stance. It is not the cleanest, truest rendering of a story. Instead, it is the negotiated outcome of a social process. And as such it is not much different (if at all) from any other social process, in how it is influenced by history, tradition, institutions, personal interactions, hopes, fears, desires.

To get to the "truth," you need multiple stories. Recall from Mara's case how the justices were struggling to divine what the medicine cartons were all about, what the strange names and figures meant. And recall from the pilot's case how he tried to show to the court that configuring the airplane for approach took more time than had been available—and nobody cared. While the professionals on trial doggedly searched for ways to get "the truth" out, it never would.

Multiple versions competed and contradicted each other, but many of them seemed equally valid. All illuminated different aspects of the case. In the Xylocard trial, the pediatrician had a point: his repeated bolus doses of Xylocard into the baby could not be judged in light of the fact that the baby was already getting 10 times the prescribed dose through her drip. He could not have known that, after all. The nurses had a point too: ordering bolus dose after bolus dose, with only worsening effects, and without ever having established a diagnosis for the baby's condition, did not make perfect sense. Settling for only one version amounts to an injustice toward the complexity of the adverse event that the nurse was on trial for.

Similarly, the captain had a point: it was the airline, its image, production pressures, and routine dispensations to as yet unlisted doctors and unqualified copilots that helped box him in. But the other side had a point too: why had this pilot not voluntarily contributed to learning and improvement after the incident?

This implies that forcing one story onto other people as if it were the true and only one (like the justice system sometimes does) is actually quite unjust. Just like the cubists try to paint multiple perspectives at the same time, a just culture always takes multiple stories into account, because

- Telling the story from one angle necessarily excludes aspects from other angles.
- No single account can claim that it, and it alone depicts the world as it is.
- Innumerable stories are possible, and, if you want to be "just," or approximate the "truth," a number are even necessary.
- Also, if you want to explore as many opportunities for safety improvement as possible, you probably want to listen to as many stories or angles as possible. The world is complex—live with it. And learn from it what you can.

A colleague in healthcare told me how he believed that some acts are objective, self-evident, or even unarguably criminal—substance above by the provider, for instance (a doctor being drunk on duty), or deliberately unsafe acts. He told me the story of some nurses who substituted diuretic tablets for pain relieving tablets as a prank to make patients demand urine bottles from the night staff. These were egregious acts, he said. Criminal acts. That could be dealt with only through discipline or other legal forms.

I am in no position to say that these things are not crimes. But what I find interesting is how we come to give the acts meanings as crimes, committed by these individuals at that moment. Seeing these acts as criminal can rule out or obscure a host of other factors that may once again trigger other people to behave similarly "criminally." When it comes to doctors deliberately murdering patients, for example, this raises a host of questions about access control to the profession (Is there a psychiatric evaluation to become a doctor? To become an airline pilot there is. Are there regular proficiency checks for doctors practicing on their own? For pilots there are). Drunk or stoned doctors raise questions about working hours (36-hour shifts, 80+-hour weeks) and the effects on their personal lives. Playing a prank on the night staff at the cost of patients raises questions about organizational staff disputes that are left unaddressed, and about the ethical awareness of the staff involved.

Yes, through the eyes of a lawyer or prosecutor, these acts may well look like crimes. The language of "crimes" is one that would seem to fit the acts above quite well. But that is not necessarily the only language in which we can talk about things such as the ones above. Or do something about them. Without the prosecutor, there would be no crime. Indeed, if given into the hands of others, these "crimes" can be constructed quite easily as different things:

- As societal or professional trade-offs (we make our doctors work long hours in part because healthcare is hugely expensive already, and we trust them to remain healthy, alert, and self-responsible once we license them)

- As managerial issues (simmering interdepartmental or cross-shift conflicts are not resolved early enough through higher-level intervention)
- As pedagogical ones (ethical training for staff)

No one can say that one interpretation is better or more "right" or just than another. But different interpretations are possible. And all interpretations have a logical repertoire of action appended to them. See only one interpretation and you may miss other important possibilities for progress on safety.

JUDICIAL PROCEEDINGS AND JUSTICE

But wait, you may say, doesn't the legal system help society understand what went wrong and why, and what we can do about it? The chances that a legal system will tease out a meaningful and just account of what happened are actually remote. It is not its charter, and even if it were, it is not particularly good at it.

Go back again to nurse Mara's case: putting all of the responsibility for the baby's death on her shoulders made no historical sense whatsoever, and was really hard to see as fair or just. Lots of other people had been involved, and she had not even administered the drug in question. The judicial proceedings in the aftermath of the baby's death, through sheer design and rules of relevancy, played down or ignored these other contributions. It ended up with an account of a complex system failure that contradicted decades of research into how such accidents actually happen.

The potential for bad outcomes lies baked into the very activity that we ask practitioners to do for us. The criminal trial of the airline captain from a case earlier in this book ("Case Study: When Does a Mistake Stop Being Honest?"), for example, found him guilty of "endangering his passengers" while flying an approach to a runway in fog. "I do that every day I fly," a colleague pilot had responded. "That's aviation."[15]

Pilots, nurses, doctors, and similar practitioners endanger the lives of this everyday as a matter of course. How something in those activities slides from normal to culpable, then, is a hugely difficult assessment, for which a judicial system often lacks the data, the education, and the expertise. The decision whether to prosecute a practitioner, then, can turn out to be quite haphazard, and the practitioner on the receiving end will likely see this as quite unjust.

In the same year that nurse Mara was first charged, more than 300 severe medication errors were reported to the country's health authority. In another study, a full 89% of responding anesthetists reported having made drug administration errors at some stage in their careers. Most had done so more than once, and 12.5% reported having actually harmed patients in this way.[88] *So why the nurse in the Xylocard case, and not one of scores of other medical practitioners who go through similar medication misadventures—all the time, everywhere?*

It is the whole point of legal proceedings to narrow in on a few acts by a few individuals or even a single individual. By its very nature, however, this clashes with what we know about accident causation in complex, dynamic systems today. Many factors, all necessary and only jointly sufficient, are needed to push a basically safe system over the edge into breakdown. Single acts by single culprits are neither necessary nor sufficient. This, logically, does not make judicial proceedings about complex events "just."

The accounts of an accident that a legal system produces can be so limited in many ways because of the way it conducts its business—among other things through

- The way judicial proceedings rationalize the search for and consideration of evidence
- How they closely script turn-taking in speech and form of expression
- How they limit what is relevant, and are institutionally constrained in their deferral to domain expertise
- How they necessarily exclude the notion of an "accident" or "human error" because there are typically no such legal concepts

This is not to deny the relevance or even authority of a legal tradition, at least not on principle. It is, rather, to see it as that: one tradition, one perspective on a case of failure. One way for which prosecutors and judges have received the power to enforce it on others, one language for describing and explaining an event, relative to a multitude of other possibilities.

Another consequence of the accountability demanded by legal systems is that it is easily perceived as illegitimate, intrusive, and ignorant. If you are held "accountable" by somebody who really does not understand the first thing about what it means to be a professional in a particular setting (a ward, a cockpit, a control room, a police beat), then you will likely see their calls for accountability as unfair, as coarse and uninformed. Indeed, as unjust. Research shows that this results in less disclosure and a polarization of positions, rather than an openness and willingness to learn for the common good.[27]

Garuda Indonesia pilot Marwoto Komar was found guilty of criminal negligence while at the controls of a Boeing 737 that slammed onto the runway at Yogyakarta Airport. The plane careered into a field and burst into flames on March 7, 2007. A majority of a panel of five judges at the Sleman District Court sentenced the pilot to two years in prison. They said there was no evidence that his plane had malfunctioned. Prosecutors had called for four years in prison. They said he ignored a series of warnings not to land as he brought the plane in far above the safe speed. Twenty-one people were killed. Relatives attending the trial said the penalty was not enough. The Indonesian Pilots Federation said it would appeal. One of the panel of five judges remarked that the sentence was not about revenge, but about the prevention of future accidents.[89]

JUDICIAL PROCEEDINGS AND SAFETY

If judicial processes in the wake of accidents can be bad for justice, what about their effects on safety? Here is a summary of some of the adverse effects.

- Judicial proceedings after an incident can **make people stop reporting incidents**. The air traffic control provider in the example in this chapter reported a 50% drop in incidents reported in the year following criminal prosecution of controllers involved in a runway incursion incident. Interestingly, the criminal prosecution does not even have to be started, let alone lead to a conviction: the threat of criminal prosecution can make people hesitant about coming forward with safety information.
- Judicial proceedings, or their possibility, can **create a climate of fear** about sharing information. It can hamper an organization's possibility to learn from its own incidents. People may even begin to tamper with safety recording devices, switching them off.
- **Judicial proceedings can interfere with regulatory work**. Some regulators, for example, have become more careful in using language such as "deviation" in their inspection reports. If it is a "deviation" that a regulator takes notice of, it is very likely a deviation from some regulation. And regulations have their basis in law. A "deviation" can then easily become a breaking of the law—a crime, rendering sources at the operator silent as a result. Regulators can become much less direct about what is wrong and needs to be done about it.
- Judicial proceedings can help **stigmatize an incident as something shameful**. Criminalizing an incident can send the message to everybody in the operational community that incidents are something professionally embarrassing, something to be avoided, and if that is not possible, to be denied, muffled, hidden.
- The stress and isolation that practitioners can feel when subject to legal charges or a trial typically **causes them to perform less well in their jobs**. And investing cognitive effort in considering how actions can get you in legal trouble detracts attention from performing quality work.[27]
- Finally, judicial proceedings in the aftermath of an accident **can impede investigatory access to information** sources, as people may become less willing to cooperate in the accident probe.[79] This could make it more difficult for investigators to get valuable information, particularly when judicial proceedings are launched at the same time as the safety investigation. There is, however, a suggestion (at least from one organization) that criminal prosecution in the aftermath of an accident does not dampen people's report willingness regarding incidents. This could point to a subtlety in how employees calibrate their defensive posture: an accident, and becoming criminally liable for one, is somehow judged to be qualitatively different from liability for incidents.

While the US National Transportation Safety Board was investigating a 1999 pipeline explosion near Bellingham, Washington, that killed three people, federal prosecutors launched their own criminal probe. They reportedly pressured employees of the pipeline operator to talk. Several invoked the US Constitution's Fifth Amendment, which protects against self-incrimination. They refused to answer questions from Safety Board investigators as well as from the police.[76]

SUMMING UP THE EVIDENCE

The cases of human error that have gone to trial so far suggest that legal proceedings—tort or criminal—in the wake of incidents or accidents could be bad for safety, and may not help in creating a just culture.

Many inside and outside professional circles see a trend toward criminalization of human error as troublesome. If justice exists to serve society, then prosecuting human error may work against that very principle. The long-term consequence for society of turning errors into crimes or culpable malpractice could be less safe systems. Criminalizing error, or pursuing tort claims, can

- Erode independent safety investigations
- Promote fear rather than mindfulness in people practicing safety-critical work
- Make organizations more careful in creating a paper trail, not necessarily more careful in doing their work
- Make work of safety regulators more difficult by stifling primary sources of information and having to package regulatory findings in a language that does not attract prosecutorial attention
- Waste money on legal processes that do not really end up contributing to justice or to safety
- Ignore needs of victims other than mere financial ones, such as apology or the recognition of having been harmed
- Discourage truth-telling and instead cultivate professional secrecy, evasion, and self-protection

If they become the main purveyor of accountability, legal systems could help create a climate in which freely telling accounts of what happened (and what to do about it) becomes difficult. There is a risk of a vicious cycle. We may end up turning increasingly to the legal system because the legal system has increasingly created a climate in which telling each other accounts openly is less and less possible. By taking over the dispensing of accountability, legal systems may slowly strangle it.

CASE STUDY

Industry Responses to Criminalization

What if you try to build a just culture in an organization that has a tendency to prosecute your employees for their involvement in incidents while at work? How does that influence what you can accomplish inside your organization? There is of course a relationship between how willing your practitioners are to report their involvement in an incident and how the outside world is likely to discover and deal with such information. Protecting your organization's data from undue outside probing is one response that gives your employees the confidence that they can report or disclose internally without fear of outside consequences. How can you do that? One response, as in the example that follows, is to do more to hide and protect your data.

Some of Canada's large commercial airlines started requiring government inspectors to sign confidentiality agreements to make sure their safety records stay private before letting them comb over company records to assess a controversial new oversight system. Over the year that followed, inspectors were going to conduct in-depth assessments of the safety management system (SMS) at Canada's large airlines, including Air Canada, WestJet, Air Transat, Porter Airlines, and Skyservice. The new safety system—a first in civil aviation—put more onus on airlines in managing safety risks in their operations and became fully phased in at Canada's large commercial carriers.[90] *Asking inspectors to sign confidentiality deals before allowing them to do the inspections suggested a climate of a lack of trust and a fear of open or honest disclosure.*

Some countries have succeeded in exempting safety data in very narrow cases from freedom-of-information legislation. The Air Law in Norway, for example, states in Article 12-24 about the "Prohibition on use as evidence in criminal proceedings" that "Information received by the investigating authority may not be used as evidence in any subsequent criminal proceedings brought against the persons who provided the evidence." Of course, this does not keep a prosecutor or judge from actually reading a final accident report (as that is accessible to all citizens), but it does prevent statements provided in good faith from being used as evidence. Similar legislation exists, though in other forms, in various countries. Many states in the United States, for example, protect safety data collected through incident reporting against access by potential claimants. Most require a subpoena or court order for release of the information.[1]

One problem with this, of course, is that such protection locks information up even for those who can rightfully claim access, and who have no vindictive intentions. Imagine a patient, for example, or a victim of a transportation accident (or the family), whose main aim is to find out something specific about what happened to their relative. The protection of reporting, in other words, can make such disclosure more difficult. So when you contemplate formally protecting reported safety information, you should carefully consider obligations to these stakeholders as well.

Another, radically different response is to disclose preemptively. Preexisting trust between stakeholders is crucial for this to work. The judiciary might be willing to let an organization handle its own incidents when it has the confidence that the organization will come to it if a case is deemed to fall outside of what it is capable of handling. This is a step that requires courage on part of the organization and its employees.

Near misses and other incidents involving aircraft at Schiphol, one of Europe's busiest airports, will from today (13 January 2015) be made public by the Dutch Air Navigation Service Provider Luchtverkeersleiding Nederland (LVNL), in a bid to reassure the public about the safety of air travel. LVNL will become the first European air traffic control provider to publish information on its website on individual incidents within days of them happening. In the UK and many other countries, these are collated and quantified to be published annually without specific information on individual incidents.

Paul Riemens, chief executive of LVNL, said the move was intended to show the public and regulators that incidents, such as when aircraft come too close to each other on a runway, were properly investigated and the lessons implemented quickly. After a year in which aviation suffered two catastrophic air accidents, the demand for increased transparency and higher safety levels had intensified. "I am convinced others will follow," he said.

The UK's Airprox Board said UK authorities would watch LVNL's initiative with interest. "We are always looking at what we can do to improve the quality of information we hold," he said. However there was a risk that making incidents known so quickly could inflame fears over air travel. "The public will see these incidents and suddenly they might think air travel is not safe, which is not the case," he said. Even after the two disasters with Malaysia Airlines and AirAsia, the international safety record in 2014 was "the safest in many years."

The Dutch move comes as Europe adopts a new way of measuring safety using a "risk analysis tool." The results will be published annually from 2016 by the European Commission. However, Mr Riemens said that given the variation of airport layouts, standard rules benchmarks could be difficult to apply. Releasing incident information would help to encourage a culture of improved safety. "People that work for the organisation know that the whole world is looking at them," he said.

LVNL has invested in information systems which will cut the time it takes to file an incident report, Mr Riemens said, which will then appear in a few hours or days. The website will show information on specific incidents, covering causes and effects as well as the measures taken to reduce the risk of similar incidents in the future. However it will not identify the airline or the individual aircraft involved.

In 2014 LVNL guided 539,548 flights in Dutch airspace. Schiphol, at Amsterdam, is Europe's fourth busiest airport with 70 per cent of passengers transferring on to flights to the rest of the world. It handled a total of 449,379 flights in 2014. Between 2007 and 2013 the number of runway incursions reported to LVNL fell from 40 to 23. Between 2011 and 2014 the number of major and serious incidents fell from 57 to 35.[91]

The key idea in the preemptive solution above is once again trust: trusting that the information that is voluntarily shared will not get used against you. That, in turn, depends on a relationship of trust between people inside of the organization, and between the organization and other stakeholders outside of it (e.g., regulators, the media, the judiciary).

Let us put the question of the protection of safety data in the context of three questions we have been asking about retributive just cultures.

1. Who in the organization or society gets to draw the line between acceptable and unacceptable behavior?
2. What and where should the role of domain expertise be in judging whether behavior is acceptable or unacceptable?
3. How protected against judicial interference are safety data (either the safety data from incidents inside of the organization or the safety data that come from formal accident investigations)?

The differences in the directions that countries or organizations or professions are taking toward just cultures come down to variations in the answers to these three questions. Some work very well, in some contexts, and others less so. Also, the list of solutions is far from exhaustive, but it could inspire others to think more critically about where they or their organization may have settled (and whether that is a good or a bad thing). The case study concludes with a joint industry response: a declaration against automatic prosecution in the wake of incidents and accidents, and suggestions for what to do instead.

Response 1: Do Nothing

This is a solution that a number of country or professions apply, perhaps because they have not yet been confronted by the consequences of judicial action against practitioners or have themselves seen the difficulty of acting in the wake of failure. This may, of course, be just a matter of time.

1. Who gets to draw the line: in most likelihood a prosecutor who has become inspired by media reports or other triggers that made her or him look more closely into an occurrence. General risk statutes, or other laws, can be used to accuse practitioners of, for example, endangering the lives of other people. Access to data to build a criminal case should be relatively easy if the country or profession has not done much or anything to prevent such judicial intrusions into their safety data. The prosecutor draws the line in the first instance, and then the judge (or jury) gets to decide.
2. The role of domain expertise is likely minimal in judging whether a line of acceptability was crossed or not. A prosecutor, for example, has none or limited domain expertise, yet she or he gets to demonstrate whether professional judgments are culpable or not. The judge is not likely to have any domain expertise either.

3. Protection of safety data is not likely to exist, and even if it does, then a country or profession that goes by local solution 1 probably has caveats in its protection so that a prosecutor can open up databases on suspicion of a crime (and the prosecutor is often the one who decides when that is the case).

Consequences

Practitioners may feel uncertain and anxious about whether "they will be next" because the rules of criminalization are left unclear and open to interpretation. Who gets criminalized for what seems random. A just culture is a long way off, and open and honest reporting could be difficult.

Response 2: The Volatile Safety Database

Some countries or professions that do not actively handle the three questions in legislation or cross-disciplinary arrangements (e.g., between their departments of transportation or health on the one hand and justice on the other) spontaneously call for the creation of a next local solution: the destroyable safety database. This means that the safety data that organizations gather themselves are stored in a form that is very easy and quick to destroy. Some safety departments have seriously considered this idea, so as to immunize themselves against prosecution. This is especially the case in countries where the organization's personnel are themselves government employees (such as some hospital workers or air traffic controllers) and can thus be forced, through various statutes and laws, to hand over anything that belongs to the state.

1. Who gets to draw the line: same as local solution 1.
2. The role of domain expertise: same as local solution 1.
3. Protection of safety data could be guaranteed, as the data will vanish when prosecutorial pressure is applied. The cost, of course, is huge: the disappearance of an organization's safety database (which can in turn violate other statutes).

Consequences

This is not really a practical solution because of the consequences of destroying a database. But that it is being considered in several countries or professions in the first place should serve as an indication of the lack of trust necessary for building a just culture. The relationship between the various stakeholders may be troubled or underdeveloped. The suspicious climate sustained by this solution will not be good for the growth of a just culture.

Response 3: Formally Investigate Beyond the Period of Limitation

In almost all countries, prosecutors have a limited number of years to investigate and prosecute crimes. In one country, the investigation of an accident took so long that the so-called period of limitation for any possible charges (seven years in this case) expired. Practitioners sighed in relief. Inspired by such apparently legitimate delaying tactics, stakeholders in other countries and professions have considered deliberately stalling an investigation so that the judiciary could not get access until the period of limitation has expired. This solution works only, of course, if the judiciary is legally limited in beginning its probe of an occurrence while the formal investigation is still ongoing. In some countries or professions this is indeed the case.

1. Who gets to draw the line: while prosecutors and judges would still be left to draw the line eventually, other parties can withhold from them both the data and the opportunity to do so.
2. The role of domain expertise is interesting in this solution, as those with more expertise of the domain (safety investigators) make a judgment of the potential culpability of the acts they are investigating. If they judge these acts to be potentially (but unjustifiably and counterproductively) culpable, they may stall an investigation until the period of limitation has expired. In this sense, investigators introduce domain expertise into the judgment of whether something is acceptable or not, but they apply this expertise in advance—anticipating how the judiciary would respond to the data they have. Investigators may of course lack the domain expertise in the legal area to really make an accurate ex ante judgment, but previous experiences or the general climate in the country or profession may give them a good basis for their conjecture.
3. Protection of safety data is pretty strong, but of course hinges on the strength of the laws and statutes that prevent the judiciary from getting access to investigation data before the period of limitation has expired. Any legal opportunities that allow the judiciary to get into the formal investigation will directly undermine this solution.

Consequences

A climate of distrust and competition between stakeholders remains strong with this solution. Rather than resolving issues on merit, stakeholders may engage in legal gaming to try to get access (or retain privileged access) to safety data for their own purposes. The climate is not encouraging for the emergence of a just culture.

Response 4: Rely on Lobbying, Prosecutorial, and Media Self-Restraint

A solution that is different from the previous ones relies almost entirely on trust between stakeholders. It has been achieved in a few countries (often after intense lobbying of lawmakers and other government officials by industry stakeholders). It has succeeded particularly in countries with strong freedom of information acts that leave their safety data exposed to both media and judiciary.

This local solution depends entirely on the extent of the trust developed and maintained, not on legal protection for any of the stakeholders. Thus, these countries typically have no protection in place for either reporters or safety data, and the judiciary has unfettered access to investigations—in principle. In practice, no prosecutor has dared to be the first to breach the trust built up. Interestingly, this solution seems to work in smaller countries that are culturally inclined toward homogeneity, trust, coherence, and social responsibility.

1. Who gets to draw the line: prosecutors would in principle get to draw the line, but they have so far not dared to draw anything. The proscription against them doing so is not a legal one, but rather cultural or political: going in and upsetting the delicate trust developed between parties is "not done" or politically not wise. But that does not mean it cannot be done. In fact, rules in countries with this solution still make exceptions for the kinds of "crimes" or "gross negligence" that prosecutors should still prosecute. The problem is of course chicken-and-egg: how is a prosecutor to find out whether a line was crossed without drawing one?
2. The role of domain expertise has been considerable in building the necessary trust between stakeholders, particularly in convincing other stakeholders (the media, the judiciary) of the value of their self-restraint, so that the entire society can benefit from safer professional systems.
3. Protection of safety data is not legally guaranteed but achieved by cultural convention and/or political pressure.

Consequences

At first sight, this solution comes across as a fraud, and as extraordinarily brittle. After all, there is nothing "on paper:" the entire contract between stakeholders to not interfere with each other's business is left to consensual agreements and trust. Practitioners may feel free to report because historically there is no threat (but can history be a guarantee for the future in this case?). On deeper inspection, though, this solution is as robust as the culture in which it is founded. And cultures can be very robust and resistant to change. This, at the same time, creates a high threshold for entry into such an arrangement: without the right cultural prerequisites, this solution may be difficult to achieve.

Response 5: Judge of Instruction

In the wake of prosecutions of practitioners that were widely seen as counterproductive to safety, some countries have moved ahead with installing a so-called judge of instruction. Such a judge functions as a go-between, before a prosecutor can actually go ahead with a case. A judge of instruction gets to determine whether a case proposed by a prosecutor should be investigated (and later go to trial). The judge of instruction, in other words, can check the prosecutor's homework and ambitions, do some investigation him- or herself, and weigh other stakeholders' interests in making the decision to go ahead with a further investigation and possible prosecution or not.

1. Who gets to draw the line: initially (and most importantly) it is the judge of instruction who gets to draw the line between acceptable and unacceptable (or: between worthy of further investigation and possible prosecution or not). Other considerations can make it into the drawing of the line too (e.g., the interests of other stakeholders).

2. The role of domain expertise is supposed to be considerable in this solution. In the solution of one country, the judge of instruction is supported by a team from the aviation industry to help determine which cases should go ahead and which not. The makeup of this team and their interaction with the judge of instruction are crucial of course. For example, if unions or professional associations are not sufficiently represented, industry representatives may decide that it is in their interest to recommend to the judge to go ahead with prosecution, as it may help protect some of their concerns.
3. Protection of safety data is managed through the judge of instruction. If prosecutors want access to safety data, they will have to go via the judge of instruction, but there are (as usual) exemptions for serious incidents and accidents.

Consequences

A judge of instruction could function as a reasonable gate-keeper—weighing the various interests before a case can even be investigated by a prosecutor. It means, though, that such a judge needs a fair representation of all stakes, and not be susceptible to asymmetric lobbying by certain parties or interests over others. As it is a rather new solution to the criminalization of human error, there are not a great deal of data yet to see whether it works well or not.

Response 6: The Prosecutor Is Part of the Regulator

A solution that takes domain expertise right up to prosecutor level is one in which the prosecutor him- or herself has a history in or affiliation with the domain, and the office of prosecutor for that particular domain is inside of the regulator.

1. Who gets to draw the line: the prosecutor gets to draw the line (to be confirmed or rejected by a judge or jury), and the prosecutor is a person from the domain and employed by a major stakeholder in the domain.
2. The role of domain expertise is considerable, as the prosecutor comes from the domain and is employed by one of its large safety stakeholders. It is thus likely that the prosecutor is better able to balance the various interests in deciding whether to draw a line, and better able to consider subtle features of the professional's performance that nondomain experts would overlook or misjudge.
3. Protection of safety data is managed as an effect of this arrangement. The regulator has interests in protecting the free flow of safety information (not only as data for its oversight, but particularly for the self-regulation of the industry it monitors).

Consequences

The integration of prosecutor and regulator can prevent unfair or inappropriate prosecution, not only because of the tight integration of domain expertise, but also because of the greater relevance of the laws or regulations that will likely be applied (as the prosecutor works for a body that makes and applies the laws for that particular domain). The risk in this solution, of course, is that the regulator itself can have played a role (e.g., insufficient oversight, or given dispensation) in the creation of an incident and can have a vested interest in the prosecution of an individual practitioner so as to downplay its own contribution. There is no immediate protection against this in this local solution, except for regulatory self-restraint and perhaps the possibility of appeals higher up in the judiciary.

Response 7: Disciplinary Rules within the Profession

A large number of professional groups (everything from accountants to physicians to hunters to professional sports players) have their own elaborate system of disciplinary rules that are meant foremost to protect the integrity of a profession. Usually, a judiciary delegates large amounts of legal authority to the boards that credibly administer these professional disciplinary rules. Professional sanctions can range from warning letters (which are not necessarily effective) to the revocation of licenses to practice. The judiciary will not normally interfere with the internal administration of justice according to these disciplinary rules. There is, however, great variation in the administration of internal professional justice and thus a variation in how much confidence a country can have in delegating it to an internal disciplinary board.

1. Who gets to draw the line: the professional's peers get to draw the line between acceptable and unacceptable. There may be pressures, of course, that go outside the actual situation considered, so as to guarantee society's (and the judiciary's!) continued trust in the system (e.g., air traffic control, healthcare) and its ability to manage and rectify itself. This may make it necessary to sometimes lay down the line more strictly so that a message of "we are doing something about our problems" clearly gets communicated to the outside—to the detriment of justice done to an individual professional. Who gets to draw the line for criminally culpable actions is an even larger problem: internal rules are not equipped to handle those, so somewhere there needs to be a possibility for judging whether outside legal action is necessary. This can be the prosecutor's initiative (but then he or she needs enough data to trigger action) or the disciplinary board (yet they likely lack the legal expertise to make that judgment).
2. The role of domain expertise is total. Domain expertise is the basis for making the judgment about the right or wrong of somebody's actions, not some externally dictated law or statute. Domain expertise is also used to consider whether to forward a case to the formal judiciary (as there will always be an escape hatch for cases of "gross negligence" and so forth). But it is at least largely domain expertise that gets to draw that line here too.
3. Protection of safety data is likely to be independent of professional disciplinary rules and would need additional legislation for formal protection. However, with a functioning (and trustworthy) internal professional disciplinary system in place, people inside a profession may feel freer to report incidents and concerns.

I came across an interesting, spontaneous variant of solving incidents and safety matters internally. An air traffic control center had essentially agreed with some of the airlines frequenting its airspace that they would send reports on near misses and other problems directly to them. This was of course a very short-loop way to learn: a problem was seen and reported directly to the air traffic control center that could do something about it. It also prevented outsiders from making their own judgments about the performance of those involved and meting out any consequences (legal, regulatory). But those other parties (e.g., the regulator) felt that accountability was being shortcut—reporting routines established earlier had ensured that reports would go through them. Lessons learned could also be relevant to the wider industry, something that was missed by this local, internal solution.

Consequences

The total integration of domain expertise in the administration of justice makes a solution based on professional disciplinary rules attractive. Not only do domain experts judge whether something is acceptable or unacceptable, but they also draw largely from the domain the "rules," written or unwritten, on which basis they make their judgment.

But there is a possible paradox in the justness of professional disciplinary rules. Because disciplinary rules exist for the maintenance of integrity of an entire profession, individual practitioners may still get "sacrificed" for that larger aim (especially to keep the system free from outside interference or undue scrutiny). To remain trustworthy in the eyes of other stakeholders, then, the disciplinary rules may have to wreak an occasional internal "injustice" so as to outwardly show that they can be trusted. This does not necessarily enhance the basis for just culture, as practitioners could still feel threatened and anxious about possible career consequences.

Aviation Safety Groups Issue Joint Resolution Condemning Criminalization of Accident Investigations[92]

Alexandria, VA—October 18, 2006—The Flight Safety Foundation (FSF), the Civil Air Navigation Services Organisation (CANSO), the Royal Aeronautical Society in England (RAeS) and the Academie Nationale de L'Air et de L'Espace (ANAE) in France today issued an unprecedented joint resolution decrying the increasing tendency of law enforcement and judicial authorities to attempt to criminalize aviation accidents, to the detriment of aviation safety.[93]

"We are increasingly alarmed that the focus of governments in the wake of accidents is to conduct lengthy, expensive, and highly disruptive criminal investigations in an attempt to exact punishment, instead of ensuring the free flow of information to understand what happened and why, and prevent recurrence of the tragedy," said Bill Voss, FSF President and CEO.

"The aviation industry is the most labor-intensive safety operation in the world, and human error is a rare but inevitable factor in the safety chain. Prosecuting basic human error is a grave mistake, as punishment should be reserved for those who are breaking the law," commented Alexander ter Kuile, Secretary General of CANSO.

"Authorities should be focusing on gathering all the facts and evidence from those involved," said Keith Mans, RAeS President, "and encouraging pilots, air traffic controllers, mechanics, design engineers, managerial officers, and safety regulatory officials to come forward and admit any mistakes without fear of retribution."

"There exist many civil and administrative mechanisms to deal with any violation of aviation standards, without resort to criminal sanctions," said Jean-Claude Buck, President of ANAE.

The safety organizations noted several recent, high-profile examples of this trend, including the ongoing investigation of the recent Embraer/Gol midair collision in Brazil, a recent French Supreme Court decision not to dismiss criminal charges stemming from the July 2000 Air France Concorde crash, an imminent verdict in the criminal trial following the 1992 Air-Inter crash in Strasbourg, France, of government, airline, and manufacturing officials, and the pursuit of criminal manslaughter charges against a number of air traffic controllers and managers of Skyguide in Switzerland in connection with the DHL/Bashkirian midair collision over southern Germany in 2002. The joint resolution makes five main points:

1. Declares that the paramount consideration in an investigation should be to determine the probable cause of the accident and contributing factors, not to criminally punish individuals.
2. Declares that, absent acts of sabotage and willful or particularly egregious reckless conduct, criminalization of an accident is not an effective deterrent or in the public's best interest.
3. Urges States to exercise far greater restraint and adopt stricter guidelines before officials initiate investigations or bring criminal prosecutions in the wake of aviation disasters.
4. Urges States to safeguard the safety investigation report and probable cause/contributing factors conclusions from premature disclosure and direct use in civil and criminal proceedings. It also criticized prosecutorial use of relatively untrained and inexperienced "experts," which can lead to "technically flawed analyses, a miscarriage of justice, and interference with official accident inquiries."
5. Urges accident investigating authorities to assert strong control over the investigation, free from undue interference from law enforcement, invite international cooperation in the investigation, conduct investigations deliberately and avoid a "rush to judgment," ensure the free flow of essential safety information, and address swiftly any acts or omissions in violation of aviation standards.

5 What Is the Right Thing to Do?

An airline safety manager told me recently of an event where an aircraft of theirs had departed from a runway after closing time of the airport, and the runway lights had been switched off. The chief pilot, on hearing about the incident, said to the safety manager: "Bring those idiots to me—now! I don't care who they are, they're dead!"

If you help run the organization, or a part of it, how would you respond to evidence of such an incident? The safety of your organization has to do with being open, with trust, with a willingness to share information about safety problems without the fear of being nailed for them. But most people also believe that the openness of a just culture is not the same as uncritical tolerance. If "everything goes," then in the end no problem may be seen anymore as safety-critical—and people will stop talking about them for that reason. It is this tension between

- Wanting everything in the open
- While not tolerating everything

This book covers how obligations to disclose are about wanting everything relevant in the open—and how a perceived lack of justice can mess that up really quickly. It covers the problems with not tolerating everything—because "everything" is not about a clear line or definition, but about who gets to decide. It covers how a just culture is about the always uneasy, but exciting melding of the two. It is exactly the tension between wanting everything in the open so that you can learn and improve, but not necessarily tolerating everything so that you can be "just."

For both retributive and restorative practices to be effective, you have to start to take an honest look at what you do in your own organization. Only this will allow you to begin building relationships and trust between the parties that matter. Even if you are torn between retributive and restorative approaches to justice in your organization, there are some things you can evaluate or accomplish before and after an incident.

DEALING WITH AN INCIDENT

Before Any Incident Has Even Happened

1. See if you can abolish financial and professional penalties in the wake of an occurrence. Suspending practitioners after an incident should be avoided too. These measures turn incidents into something shameful. If your organization has these kinds of rules in place, you can lose out on much valuable safety information.
2. Explore having a staff safety department, not part of the line organization, that deals with incidents. The direct manager (supervisor) of the practitioner

should not necessarily be the one who is the first to deal with that practitioner in the wake of an incident (other than perhaps relieving him or her temporarily to deal with the stress and aftermath). Aim to decouple an incident from what may look like a performance review of the practitioner involved. Any retraining of the practitioner involved in the incident will quickly be seen as punishment (and its effects are debatable), so this should be done with utmost care and only as a last resort. In fact, whether a practitioner should undergo retraining, for example, is something that should be discussed not only with the practitioner in question (rather than just handed down from above), but also checked with a group of peers who can consider the wider implications of such a measure in the wake of an incident (e.g., on the reputation of that practitioner, but also on the way incidents will be seen and treated by colleagues as a result).
3. Be sure that practitioners know their rights and duties in relation to incidents. Make clear what can (and typically does) happen in the wake of an incident. One union had prepared little credit-sized cards on which it had printed the practitioner's rights and duties in the wake of an occurrence (e.g., to whom they were obliged to speak, such as investigators) and to whom not to speak (e.g., the media). Even in a climate of anxiety and uncertainty about a judiciary's position on occurrences, such knowledge about rights and duties will give your practitioners some anchor, some modicum of certainty about what may happen. At the very least this will prevent them from withholding valuable incident information because of misguided fears or anxieties.
4. Start with building a just culture at the very beginning, during basic education and training for the profession. Make trainees aware of the importance of reporting incidents for a learning culture, and get them to see that incidents are not something individual or shameful but systemic information about and for the entire organization. Convince new practitioners that the difference between a safe and an unsafe organization lies not in how many incidents it has, but in how honestly it deals with the incidents that it has.
5. Implement or review the effectiveness of any debriefing programs or critical incident/stress management programs you may have in place to help practitioners after incidents (and if you don't have any in place, consider building such programs). Such debriefings and support form a crucial ingredient in helping practitioners see that incidents are "normal," that they can help the organization get better, and that they can happen to everybody. Empowering and involving the practitioner him- or herself in the aftermath of an incident is the best way to maintain morale, maximize learning, and reinforce the basis for a just culture.

After an Incident Has Happened

1. Do what you can to make sure the incident is not seen as a failure or a crisis, neither by management nor by colleagues. An incident is a free lesson, a great opportunity to focus attention and to learn collectively. Don't ask who is responsible, but ask what is responsible for producing the incident.
2. Monitor and try to prevent stigmatization of the practitioners involved in an incident. They should not be seen as failures or as liabilities to work with

by their colleagues. This is devastating not only for them, but also for every practitioner and by extension the organization. Reintegrate practitioners into the operation smoothly and sensitively, being aware of the possibility for stigmatization by their own colleagues.

3. Conduct a learning review of the event, not necessarily an investigation. A learning review asks not *who* was responsible for causing the incident, but *what* was responsible. It tries to control the hindsight bias. It focuses on forward-looking accountability rather than backward-looking accountability (see later for more on these topics).
4. If your organization pursues a retributive path after an incident, at least make sure that the substantive and procedural assurances from Chapter 1 are in place that can make such a process anywhere near "just." Also, ask yourself whether this response address harms, needs, *and* deeper causes, and whether it is adequately victim oriented (including both first and second victims) and respectful to all parties and stakeholders involved in the aftermath of the incident.
5. If your organization wishes to pursue a restorative path instead, then be sure to ask who is hurt both inside and outside your organization, what their needs are, and to what extent you have the obligation to meet any of those needs. If not you, then who does have that obligation? Your organization will also have needs (this may range from public image to business continuity). You will need to find out what serves those needs best, and how you can avoid them from clashing with other needs (e.g., those of the first and second victims) by making the just culture process both inclusive and collaborative.

Ivan "Pup" Pupulidy, one of my students and an officer in the US Forest Service, introduced the idea of a learning review to his community. If we don't find fault after an incident, he asked, then what do we do? The aim of the learning review, he said, is to understand the rationale for the actions and decisions involved in the incident and then, if possible, to learn from them. Achieving this goal, he argued, requires a deliberate effort to place decisions and actions in the detailed context in which they occurred, including an understanding of the pressures faced by all those involved in the incident.

People involved in the review were asked to suspend judgment, from the perspective of hindsight, and instead engage as active participants in the process of learning from the event. By coming to understand why it made sense for the people involved, at all levels of the organization, reviewers began to understand what makes the system brittle and what might make it resilient.

The learning review, he said, will not draw conclusions in the traditional way that reports have done in the past. Conclusions can sometimes close the door on learning, by suggesting that all information has been found and judgments can be made. Judgment, however, is always biased from the perspective of the writer and/or reader. So this review allows sometimes unresolved concepts to emerge, allowing tension to be created for the reader. This tension can inspire dialogue in the community and the organization, and encourages sensemaking around the presented concepts. This way, the learning review can become a living process. Introducing one of the learning reviews, he said, "This review will answer some questions, but it is likely to raise others. It is designed to do just that, which makes it different from traditional reports. Look inside the cover of this Review for answers, but more importantly, look inside yourself."[94]

There might be questions from the regulator who is watching over you. It can sometimes be difficult to persuade a regulator that you are doing something about the problem if you do not show a strong retributive response. And indeed, this may also have implications for your own position or survival as organizational manager. I have seen that it takes courage to convince the regulator, colleagues, a board, and indeed other parties in society, that you are doing *a lot* when you respond restoratively. In particular, you help the organization learn and hopefully prevent similar events in the future. You can ensure that the stories, the accounts surrounding the tragic event, are preserved and shared. You can take on board the recommendations they imply. You can create the conditions for a culture where your people feel free to share safety-critical information with you. You show you trust them with safety-critical work, and they trust you to treat them fairly and humanely when things go wrong.

An operations manager of a mine recently asked me whether I thought it would be just to fire employees who show up for work while drunk or drugged. This is precisely what his mine does. The employee is fired, no questions asked, no appeal offered.

In a raw, narrow sense, that could be seen as just, I told him. After all, if this person is going to operate heavy equipment that can harm other workers, you want the person to be sober.

But, I said, perceptions of injustice may start to seep in if that is all that you do. Consider the location of your mine, I said. It is in the absolute middle of nowhere, a hot, dusty place. You have fly-in, fly-out workers, like on an oil rig. When out there, the guys (mostly guys) are isolated. They have nothing much for entertainment, very little in the way of social support except each other.

I am not making excuses for somebody who shows up drunk or drugged, I said, but what you want to consider is opening up a parallel inquiry into the conditions that make it more likely for people to do so. What is it in the situation in which you configure them that makes them vulnerable? If you don't do that, and just fire your people, then other workers might begin to wonder about your understanding of their conditions, your interest in them, and indeed whether what you are doing is just after all. And you might well harm your own organization, I said. You are going to need those people. They are hard to replace. And you do not want your mine to get a reputation that keeps people from wanting to work there.

Finally, I said, if you want to keep your instant-firing policy in place, you want to make sure that the decisions about firings have a good foundation in your workforce. That it is not just you who decides. Make sure there is a constituency for the policy, involve your workers in its language, its application, and the judgments that arise from it. They are the ones, after all, who are going to have to work, or not, with that person. If they feel that judgment is going to be passed over their heads, without any involvement, workers may start covering for each other—helping colleagues hide the evidence of drunkenness or drugs. As a manager, you won't know about it, but the safety of your operation is going to be hollowed out from the inside out.

NOT INDIVIDUALS *OR* SYSTEMS, BUT INDIVIDUALS *IN* SYSTEMS

For some, this might still leave the question from the preface unanswered. Can people in your organization simply blame the system when things go wrong? To many, this seems like a cop-out, like an excuse to get defective or responsible practitioners off the hook. Of course we should look at the system in which people work, and improve it to the best of our ability. But safety-critical work is ultimately channeled through relationships between human beings (such as in healthcare), or through direct contact of some people with the risky technology. At this sharp end, there is almost always a discretionary space into which no system improvement can completely reach. Rather than individuals *versus* systems, we should begin to understand the relationships and roles of individuals *in* systems.[17]

A Discretionary Space for Personal Accountability

A system creates all kinds of opportunities for action. And it also constrains people in many ways. Beyond these opportunities and constraints, we could argue that there remains a discretionary space, a space that can be filled only by an individual care-giving or technology-operating human. This is a final space in which a system really does leave people freedom of choice (to launch or not, to go to open surgery or not, to fire or not, to continue an approach or not). It is a space filled with ambiguity, uncertainty, and moral choices.

Systems cannot substitute the responsibility borne by individuals within that space. Individuals who work in those systems would not even want their responsibility to be taken away by the system entirely. The freedom (and the concomitant responsibility) that is left for them is what makes them and their work human, meaningful, a source of pride. But systems can do two things.

1. One is to be as clear as possible about where that discretionary space begins and ends. Not giving practitioners sufficient authority to decide on courses of action (such as in many managed care systems), but demanding that they be held accountable for the consequences anyway, creates impossible and unfair double binds. Such double binds effectively shrink the discretionary space before action, but open it wide after any bad consequences of action become apparent (then it was suddenly the physician's responsibility after all).

   The same goes when asking dispensation for an unqualified crewmember to proceed with an instrument approach in the November Oscar 747 case described at the beginning of this book. The system is clear in its routine expectation that a commander will ask such dispensation. And if all goes well, no questions will be raised. But if problems occur on the approach, the request for dispensation suddenly becomes the commander's full responsibility. Such vagueness of where the borders of the discretionary space lie is typical, but it is unfair and unreasonable.
2. The other thing a system can do is decide how it will motivate people to conscientiously carry out their responsibilities inside of that discretionary space.

> Is the source for that motivation going to be fear or empowerment? Anxiety or involvement? "There has to be some fear that not doing one's job correctly could lead to prosecution," said an influential commentator in 2000. Indeed, prosecution presumes that the conscientious discharge of personal responsibility comes from fear of the consequences of not doing so. But neither civil litigation nor criminal prosecution work as a deterrent against human error. Instead, anxiety created by such accountability leads, for example, to defensive medicine, not high-quality care, and even to a greater likelihood of subsequent incidents.[60] The anxiety and stress generated by such accountability adds attentional burdens and distracts from conscientious discharge of the main safety-critical task.[27] Rather than making people afraid, systems should make people participants in change and improvement. There is evidence that empowering people to affect their work conditions, to involve them in the outlines and content of that discretionary space, most actively promotes their willingness to shoulder their responsibilities inside of it.[30]

Haavi Morreim reports a case in which an anesthesiologist, during surgery, reached into a drawer that contained two vials, sitting side by side.[95] *Both vials had yellow labels and yellow caps. One, however, had a paralytic agent and the other a reversal agent to be used later, when paralysis was no longer needed. At the beginning of the procedure, the anesthesiologist administered the paralyzing agent, as per intention. But toward the end, he grabbed the wrong vial, administering additional paralytic instead of its reversal agent. There was no bad outcome in this case. But when he discussed the event with his colleagues, it turned out that this had happened to them too, and that they were all quite aware of the enormous potential for confusion. All knew about the hazard, but none had spoken out about it.*

The question is of course why no anesthesiologist had flagged this problem before. Anxiety about the consequences of talking about possible failures cannot be excluded: it has squelched safety information before.

Even more intriguing is the possibility that there is no climate in which practitioners feel they can meaningfully contribute to the context in which they work. Those who work on the safety-critical sharp end every day, in other words, did not feel they had a channel through which to push their ideas for improvement. I was reminded of one worker who told me that she was really happy with her hospital management's open-door policy. But whenever she went through that open door, the boss was never there.

Do we really think we can prevent anesthesiologists from grabbing a wrong vial by making them afraid of the consequences if they do? Or do we want to prevent them from grabbing a wrong vial by inviting them to come forward with information about that vulnerability, and giving us the opportunity to help do something more systemic about the problem?

Blame-Free Is Not Accountability-Free

Holding people accountable and blaming people are two quite different things. Blaming people may in fact make them less accountable: they will tell fewer accounts, they may feel less compelled to have their voice heard, to participate in improvement efforts. This also means that blame-free or no-fault systems are not accountability-free systems. On the contrary: such systems want to open up the ability for people to hold their account, so that everybody can respond and take responsibility for doing something about the problem.

Equating blame-free systems with an absence of personal accountability is short-sighted and not very constructive. Blame-free means blame-free, not accountability-free. The question is not whether we want practitioners to skirt personal accountability. Few practitioners do. The question is whether we want to fool ourselves that we can meaningfully wring such accountability out of practitioners by blaming them, suing them, or putting them on trial. No single piece of evidence so far seems to demonstrate that we can. We can create such accountability not by blaming people, but by getting people actively involved in the creation of a better system to work in. Most practitioners will relish such responsibility, just as most practitioners often despair at the lack of opportunity to really influence their workplace and its preconditions for the better.

John Allspaw, one of my students and a software engineer, had this way of explaining the idea of "blameless" postmortem to his community.[96] *It means investigating mistakes in a way that focuses on the situational aspects and the decision-making process of individuals proximate to the failure. Having a "blameless" postmortem process means that engineers whose actions have contributed to an accident can give a detailed account of*

- *What actions they took at what time*
- *What effects they observed*
- *Expectations they had*
- *Assumptions they had made*
- *Their understanding of timeline of events as they occurred*

...and that they can give this detailed account without fear of punishment or retribution.

Why shouldn't they be punished or reprimanded? Because an engineer who thinks he or she is going to be reprimanded is disincentivized to give the details necessary to get an understanding of the mechanism and operation of the failure. This all but guarantees that it will repeat itself—if not with the original engineer, then with another one in the future.

If we go with "blame" as the predominant approach, then we're implicitly accepting that deterrence is how organizations become safer. This is founded in the belief that individuals, not situations, cause errors. It's also aligned with the idea there has to be some fear that not doing one's job correctly could lead to punishment. Supposedly, the fear of punishment will motivate people to act correctly in the future. This cycle of name/blame/shame can be looked at like this:

1. *Engineer takes action and contributes to a failure or incident.*
2. *Engineer is punished, shamed, blamed, or retrained.*
3. *Trust is reduced between engineers on the ground (the "sharp end") and management (the "blunt end") looking for someone to scapegoat.*

4. *Engineers become silent on details about actions/situations/observations, resulting in "cover-your-ass" engineering (from fear of punishment).*
5. *Management becomes less aware and informed on how work is being performed day to day, and engineers become less educated on lurking or latent conditions for failure due to silence mentioned in step 4.*
6. *Errors are more likely; latent conditions can't be identified due to step 5.*
7. *Repeat from step 1.*

We need to avoid this cycle. We want the engineer who has made an error to give details about why (either explicitly or implicitly) he or she did what they did; why the action made sense to them at the time. The action made sense to the person at the time they took it, because if it hadn't made sense to them at the time, they wouldn't have taken the action in the first place. Only by constantly seeking out its vulnerabilities can organizations enhance safety.

A funny thing happens when engineers make mistakes and feel safe when giving details about it: they are not only willing to be held accountable; they are also enthusiastic in helping the rest of the company avoid the same error in the future. They are, after all, the most expert in their own error. They ought to be heavily involved in coming up with remediation items.

So technically, engineers are not at all "off the hook" with a blameless postmortem process. They are very much on the hook for helping us become safer and more resilient. Most engineers find this idea of making things better for others a worthwhile exercise.

So what do we do to enable a "just culture"?

- *We encourage learning by having blameless postmortems.*
- *The goal is to understand how an accident could have happened, in order to better equip ourselves to prevent it from happening in the future.*
- *Instead of punishing engineers, we allow them to give detailed accounts of their contributions to failures. We gather details from multiple perspectives.*
- *We accept that the hindsight bias will cloud our assessment of past events, and we work hard to eliminate it.*
- *We make sure that the organization understands how work actually gets done (as opposed to how they imagine it gets done, via charts and procedures).*
- *Operational people inform the organization where the line is between appropriate and inappropriate behavior. This isn't something that managers or others can come up with on their own.*

Failure happens. Let's take a hard look at how the accident actually happened, treat the engineers involved with respect, and learn from the event.

FORWARD-LOOKING ACCOUNTABILITY

"*He or she has taken responsibility, and resigned.*"

We often say this in the same sentence. We may have come to believe that quitting and taking responsibility are the same thing. Sure, they can be. But they don't have to be. In fact, holding people accountable may be exactly what we are *not* doing when we allow them to step down and leave a mess behind.

Accountability is often only backward-looking. This is the kind of accountability in trials or lawsuits, in dismissals, demotions, or suspensions. Such accountability tries to find a bad apple, somebody to blame for the mess. It is the kind of accountability that feeds a press (or politicians, or perhaps even a company's board), who may eagerly be awaiting signs that "you are doing something about the problem." But for you and your organization, such backward-looking accountability could be pretty useless or even harmful—other than getting somebody's hot breath off of your neck.

Instead, you could see accountability as looking ahead. Stories of failure that *both* respond to calls for accountability and allow people and organizations to learn and move forward, are essentially about looking ahead. In those stories, accountability is something that brings information about needed improvements to people or groups who can do something about it. There, accountability is something that allows people and their organization to invest resources in improvements that have a safety dividend, rather than deflecting resources into legal protection and limiting liability.

Virginia Sharpe, a philosopher and clinical ethicist who has studied the problem of medical harm for many years, has captured these dual demands in what she calls "forward-looking accountability."[1] Accountability that is backward-looking (often the kind in trials or lawsuits) might try to find a scapegoat, to blame and shame an individual for messing up. But accountability is also about looking ahead. Not only should accountability acknowledge the mistake and the harm resulting from it. It should also lay out the opportunities (and responsibilities!) for making changes so that the probability of recurrence is reduced. In the words of Sharpe,

> The forward-looking or prospective sense of responsibility is linked to goal-setting and moral deliberation. Responsibility in this sense is about the particular roles that a person may occupy, the obligations they entail, and how those obligations are best fulfilled. But whereas responsibility in the retrospective sense focuses on outcomes, prospective responsibility is oriented to the deliberative and practical processes involved in setting and meeting goals.
>
> Currently, the dominant view of responsibility is compensation to harmed parties and deterrence of further malpractice. Responsibility in this context is retrospective; its point is the assignment of blame. A systems approach to error emphasizes responsibility in the prospective sense. It is taken for granted that errors will occur in complex, high-risk environments, and participants in that system are responsible for active, committed attention to that fact. Responsibility takes the form of preventive steps to design for safety, to improve on poor system design, to provide information about potential problems, to investigate causes, and to create an environment where it is safe to discuss and analyze error.[1]

An explosion occurred at a Texas oil refinery in March 2005, as an octane-boosting unit overflowed when it was being restarted. Gasoline vapors seeped into an inadequate vent system and ignited in a blast that was felt five miles away. The explosion killed 15 people. An internal company study into the accident found that four of the company's US executives should be fired for failing to prevent the explosion, and that even the company's global refinery chief had failed to heed serious warning signals. The company's "management was ultimately responsible for assuring the appropriate priorities were in place, adequate resources were provided, and clear accountabilities were established for the safe operation of the refinery," said the lead company investigator.

Corporate budget cuts had compromised worker safety at the plant, an earlier report had found, and the refinery had had to pay a record fine for worker safety violations at its site. A safety culture that "seemed to ignore risk, tolerated noncompliance and accepted incompetence" was determined as a root cause of the accident. The global refinery chief should have faced and communicated "the brutal facts that fundamentally, the refinery was unsafe and it was a major risk to continue operating it as such."[97]

Calls for accountability are important. And responding adequately to them is too. Sending the responsible people away is of course one response. But, remember from Chapter 1, that calls for accountability are in essence about relationships and trust. This means that breaking those relationships by getting rid of a few people (even if they are in positions of greater responsibility) may not be seen as an adequate response. Nor is it necessarily the most fruitful way for an organization to incorporate lessons about failure into what it knows about itself, into how it should deal with such vulnerabilities in the future.

Ask *What* Is Responsible, Not *Who* Is Responsible

The question that drives safety work in a just culture is not *who* is responsible for failure. Rather, it asks what is responsible for things going wrong. *What* is the set of engineered and organized circumstances that is responsible for putting people in a position where they end up doing things that go wrong?

Shortly after midnight on June 21, 1964, James Chaney, Michael Schwerner, and Andrew Goodman were murdered by a group of White Citizens' Council and Ku Klux Klan members in Mississippi. The three young civil rights activists had been in the state to help black Americans register to vote. A Neshoba County deputy sheriff, Cecil Price, stopped the three men on a tip from other white activists in Meridian, Mississippi, jailed them, and instructed his secretary to keep quiet about their incarceration. Meanwhile he notified his Klan associates, who assembled and planned how to kill the three civil rights workers.

With a fine of $20, the three men were ordered to leave the county. Price followed them to the edge of town, but pulled them over again and held them until the Klan arrived. They were taken to an isolated spot where Chaney, a black man, was mutilated and all three were shot dead. A local minister was part of the Klan group that attacked them.

The bodies were not located until weeks later, and the outrage over their killings helped bring about the passage of the 1964 Civil Rights Act. Commenting on the crimes not long after, Martin Luther King, Jr. urged people to ask not who was responsible, but what was responsible for the deaths. What was the mix of hatred, of discrimination, of bigotry and intolerance, of fear, ratified in how counties were run, in how politics was done, in how laws were written and selectively applied? It was that mix that drove men to see their acts as legitimate, as necessary for their survival and continued supremacy. King's was a system-level appeal avant-la-lettre, a quest to go up and out in seeking an understanding of why such evil could happen, rather than a down-and-in hunt for a few bad Klan apples.

In the search for the three young men (two of them white), at least seven bodies of blacks turned up. Many of them had been missing for months, without much action or even attention from authorities. Missing, murdered blacks were the norm. Similar norms or fixed ideas prevailed. An earlier trial had hung because one tormented Mississippi jury member could not stomach declaring the minister guilty.

It took decades to convict Ray Killen, one of the Klan members involved. Opposing his three consecutive 20-year sentences, Killen argued in 2005 that no jury of his peers at the time would have found him guilty. He was probably right. The operation of the institution of justice might not have led to justice. In fact, it took extrajudicial action to achieve justice. A mafia member (of the Colombo crime family) was allegedly recruited by the FBI to help find the bodies. He threatened a Klansman by putting a gun in his mouth, forcing him to reveal the location. Illegal, but quite just, according to many.

Organizations concerned with building a just culture do not normally struggle with forces as deep, pervasive, and dark as those that killed Chaney, Schwerner, and Goodman. They will not be asked to make sense of the behaviors of Klansmen, nor take a position on whether it invites sanction or not. Yet the question that King raised—ask not who is responsible, but what is responsible—rings as relevant for us now as it did then. The aim of safety work is not to judge people for not doing things safely, but to try to understand why it made sense for people to do what they did—against the background of their engineered and psychological work environment. If it made sense to them, it will for others too. Merely judging them for doing something undesirable is going to pass over that much broader lesson, over the thing that your organization needs to do to learn and improve.

Offloading a failure onto a few individuals is not usually going to get you very far. The conclusion drawn from most incidents and accidents in aviation is that everybody and everything contributes in a small way. These small events and contributions can combine to create unfortunate and unintended outcomes. People do not come to work to do a bad job. Like King, people concerned with safety must try to understand not who is responsible for an error, but *what* is responsible. What set of circumstances, events, and equipment put people in a position where an error became more likely, and its discovery and recovery are less likely? The aim is to try to explain why well-intended people can act mistakenly, without necessarily bad intentions, and without purposefully disregarding their duties or safety.

WHAT IS THE RIGHT THING TO DO?

The dead girl was wrapped in a shower curtain.

When the police found the package in the trunk of her stepfather's car, they noticed the little girl had a rag stuffed down her throat, secured in place with a bandage around her head. It turned out that the rag had been put there by her mother, who, days before, had shoved the girl under a bed and left her. Under that bed, she had died alone. The rag would have kept her from crying—crying from abandonment, fear, hunger. At three years of age, her body weighed 10 kg, or about 20 lb. Her body was now on its final journey, to be dumped in a wood.

How can a society accomplish justice in the aftermath of something like that? What is the right thing to do?

The mother was charged and convicted and got six years in jail. That would be just to many people. To them, it would be the right thing to do. Some might say that the sentence was too short. The mother was also forced into treatment. Not right, some would say, not deserved. Very appropriate, and very smart and just, others would say.

But the prosecution was not satisfied. Not yet. They found another contributing party and produced a charge of manslaughter in the second degree.

The stepfather, you'd think. Aiding, abetting, driving a car with a body in the trunk? That would be him.

Except it wasn't.

The charge was leveled against the social worker who replaced the original worker tasked with looking after this family. The alleged failure of the replacement social worker was that she did not pick up on signals of the child's neglect. Here, in short, was the case. The family had been troubled from the very start. When the girl was one year old, the state had taken her out of the mother's care because of signs of abuse, only to be returned to the family not much later. The paperwork that would have testified as to the family fulfilling the conditions for the child's return, however, was lost, or never produced. The child protection council was not notified either.

The social worker visited the family three times, and found little to report. After a while, she went on sick leave. It took months before a replacement was found. The replacement worker drew up a plan for the mother. It specified, among other things, what to give to a toddler to eat, when, how often, and other basic things related to child care and hygiene. The mother never really managed. The girl started falling behind in language, and started to look a little blue.

And then, one day, she was dead.

WHAT CAN ETHICS TELL YOU?

Was charging the replacement social worker with manslaughter the right thing to do? This is an ethical question. It is a question about our values, about what we consider to be right or wrong. That finding an answer might be hard, however, does not mean that the question resists systematic reflection. Ethics, as a branch of philosophy, offers that sort of systematic reflection. When ethical thinking departs from the merely descriptive and becomes prescriptive, it offers systematic ways of considering what people ought to do in a particular situation. The different ethical approaches offer different ways of mapping the same moral terrain. Here are the approaches that are explained a bit more below:

- Virtue ethics
- Duty ethics

- Contract ethics
- Utilitarianism
- Consequence ethics
- Golden rule ethics

These approaches in themselves do not offer easily applicable solutions perhaps, and any brief treatment of what they might offer is likely unfair. But here is some guidance you can start with.

Virtue Ethics

Virtue ethics comes in part from the ideas of Aristotle and also Confucius, and asks what it takes to be a good or virtuous person. You can ask that of yourself as you ponder what to do in the wake of an incident in your organization. You can also ask it of the practitioner who was involved. Virtues are reliable habits that you engrave into your identity. They are considered to be a constant of your character, rather than driven by the different roles you might have. Virtues are typically surrounded by vices on either side. The virtue of courage, for example, has the vice of cowardice on one side and rashness on the other side. In the wake of an incident, you want to be neither: you probably want to be courageous—to disclose if you were involved in an incident, and to be restorative rather than punitive when your job it is to respond to that incident. In this, you want the virtue of being temperate, but not the vices on either side: being insensible or indulgent, respectively. Aristotle was inspired by people who possessed what he called *phronesis*, the sort of practical wisdom or on-the-spot ability to see what is good or virtuous in any situation and how to achieve it. The way to become virtuous, he proposed, was to be driven to take on that goal, be inspired by people whose traits you would like to have, and then practice them yourself. The right thing to do, according to virtue ethics, is that which is in line with the virtues you wish to have, and you wish others to have. That still leaves it a bit vague in the case described earlier, of course. Which is indeed one of the critiques of virtue ethics. Not only is its guidance rather underspecified; it is hard to know which virtues are inherently "right" or "good."

Duty Ethics

Duty ethics is the ethics of principle. Principles are different from rules. Rules are followed because they are externally imposed and policed. Principles are something you can make a part of yourself, particularly in relation to a professional duty. So this kind of ethics studies the nature of that duty, and the obligations and constraints that come with it. To Kant, an important moral philosopher, such obligations were not conditional, not negotiable.

Professions where (potentially risky) decisions about the lives of other people get taken come with a duty of care (which, incidentally, does get reinforced through laws relating to such a duty of care). The relationship of trust between professional and client (patient, family, passenger, child) is often called a fiduciary relationship. The client has comparatively limited knowledge and power to influence what the professional might do or decide. The relationship, and people's willingness to engage in it, is founded on the trust that the professional knows what she or he is doing, and does the best for the person in her or his care. This is where deontology might suggest that going after the replacement social worker is ethical, is the right thing to do. She did not live up to her duty of care. She violated the fiduciary relationship. She knew what the child and mother needed, or should have known. And she should have ensured that this was leading to a safe situation for the child, not a lethal one.

But, of course, things are not as simple as that. The fiduciary relationship is also founded on the belief that the professional will do everything in the best interest of the client in front of her or him. When meeting with a client—a family, a patient—*nothing* in the world should be more important than the client seen there and then. The financial bottom line is not more important, nor is the clock, nor the next client waiting to be seen. The duty to do the best for the current client overrules them all.

But that works only in an ideal world. Giving all the time and resources to one family (living up maximally to the duty ethic relative to that client) takes away time and resources from others. This militates against the ability to live up to the duty ethic with those other clients. It creates a classic goal conflict, or ethical conflict even, for social workers (as it does for many physicians). And it could be argued that most families or patients deserve or require more time than is accorded them. This is, in most Western countries, a structural constraint for services such as social work, state family support, child protection, or healthcare. They are always under pressure of limited resources—not enough money, not enough people (remember it took months to find a replacement in the little girl's case), not enough time. And always more families or patients to be seen, waiting for help, attention.

So part of being a good professional, of living up to the duty ethic, is making sure that *all* families get the best care you can give them. That, of course, starts to sound like utilitarianism (see later): the best to the most, the greatest good to the greatest number. A good duty ethic under limited resources and goal conflicts, then, means being a good utilitarian. It means juggling time and resources in a way that gets the most to the most families. But of course this militates against a more pure reading of duty ethic—that nothing is more important than the family seen there and then. There is no hope that such an ethical conflict can ever be resolved. It is felt by most social workers, and most healthcare workers, every day, all over the world. Organizations that employ or deploy such professionals often do little to encourage serious reflection over moral conflict, nor do they help their people manage it. The conflict simply gets pushed down into the workday, to be sorted out at the sharp end, on the go, as a supervisor draws up the schedules, as a social worker hurries from one family to the next.

This complicates any judgment about whether somebody lived up to professional duty. If we want to come to a fair judgment of whether pursuing the replacement social worker is the right thing to do, then there is a lot more we need to look at. Just considering the dead girl and connecting that, in hindsight, to the (now so obvious) signals of neglect that the social worker should (now so obviously) have picked up and acted on, is not going to be enough. What was the case load for this worker? What were the handover procedures when getting cases from the previous worker? How did signals of neglect come in over time, and how did they compare to the perceived criticality of the signals coming from other families in the care of this worker? Who made the schedules and what rationale were they based on? And we could go on. How was social work funded and staffed and organized in this state? Whether prosecuting the social worker is the right thing to do would depend on a careful collage of answers to all of those questions, and probably more.

A colleague recently received a phone call from a hospital vice president. A child had died a few days before from a 10-fold chemotherapy overdose in their pediatric oncology unit. He led the investigation of this tragedy and found a number of issues in processes, admixture formulation practices, and problematic new pharmacy technology that aligned to bring about the death of this child. The family was devastated. Everyone involved in the care of the child was devastated. Suddenly and lethally, doing what they usually did to create and administer chemotherapy admixtures had not worked as intended. The introduction of the new pharmacy device was deemed a substantial factor—it had replaced familiar technology "on the fly" and this was one of the first uses.

The vice president said that he did not believe any of the personnel involved should be punished. Yet, despite his organization's publicly announced plan to develop a just culture, the chief executive, chief medical officer, and human resources director insisted on firing the two pharmacists involved in the formulation of the admixture and the nurse who had administered the medication. There was absolutely no way the nurse could have known that the content of the IV bag was not as labeled. The impetus for dismissal actually came from their consulting ethicist, who also happened to be a lawyer. He identified the child's death as evidence of a breach of "duty ethic" and hence a breach of legal duty—he deemed these three people unequivocally negligent.

Contract Ethics

Going back to the original example that opened this chapter, contract ethics considers the contractual arrangement under which the social worker was hired and under which she worked with and for families. In the aftermath of the child's death, the answer to the question, "What is the right thing to do"? would be driven by what is in those contracts (though contracts do not always have to be written down). The problem for contract ethics is ensuring that both sides live up to the agreement. Contract ethics proposes that we may need an arbiter, a sovereign, who can help settle disputes and rule in cases where contracts have been breached. For this to work, however, people need to give over, or alienate their rights to that arbiter or sovereign. In other words, they have to hand their rights in, without any expectation of getting them back. That takes a lot of trust. The social worker, for example, as a member of the society in which this case played out, had little choice but to submit to the final judgment, as did Mara the nurse, or the captain of the November Oscar 747. Alienating your rights under a contract (legal, societal, or an employment contract) requires quite a bit of trust: trust that you will be treated fairly, and that your rights will not be trampled in favor of other members of society. And if we think this is difficult in a democracy (where people get to choose their head of state), how does this work in a corporation that does not have to be, and hardly ever is, democratic? This is where, if you are making decisions inside of such an organization or institution, you want to be guided by other ethical thinking than just contract ethics.

UTILITARIANISM

According to utilitarianism, the ethical or right thing to do is that which produces the greatest good for the greatest number. Getting rid of an unsafe person (removing a social worker who does not pick up signals of neglect) could then qualify as ethical. The benefit to families, to children, to co-workers, and the organization is greater than any cost. In fact, the cost is borne mostly or exclusively by the individual who is removed and charged. All the possible benefits go to a lot of people, the cost goes to one. It could be argued that an even greater good goes to an even greater number here—the society surrounding this family and their state caretakers. They receive the good that those complicit in the death of the little girl get punished. Getting rid of a bad apple, a deficient worker, harms virtually no one and benefits many people. In fact, any harm is inflicted only on the person or party who might deserve it anyway. So utilitarianism could perhaps argue that this would be the right thing to do. The critique, of course, is that utility and justice do not overlap, as they indeed don't in this case. The option that gives the highest utility (the greatest good for the greatest number) is also one that imposes a grave injustice on the social worker. Indeed, one wonders whether punishing the social worker actually creates the greatest good possible. We can evaluate that when we consider the consequences of such punishment a bit more.

CONSEQUENCE ETHICS

Consequence ethics is a school of ethical thinking that also includes utilitarianism. What are the consequences of charging the replacement social worker with manslaughter? Of course, there are all kinds of consequences. Not in the least for the social worker herself. Chapter 4 considers the consequences for the "second victim" in more detail. But what matters here are the consequences for the profession, and for the people in its care: children like the one who died. One predictable consequence is this: prosecution of the social worker is likely to tell her colleagues that they should look harder and intervene more aggressively—or else.

And so they did. The very next year, the number of children taken out of their families' care in this state doubled. Only very weak signals, or mere hints of trouble, would be necessary for a social worker to decide to intervene. The cost of missing signals is simply too large. But that sort of response has consequences too: the cost gets displaced. It gets moved around the system and part of it may well end up on the heads of the most vulnerable ones. Because where do those children go? While in the care of the state, many would go to foster families or other temporary solutions. In many countries appropriate foster families are difficult to find, even with normal case loads. Doubling the number of children from one year to the next can lead to a lowering of standards for admitting foster families. This can have consequences for the safety and security of the children in question.

And there are more consequences. Doubling the number of cases from one year to the next will lead to a doubling or at least an increase of the paperwork and in the supervisory and organizational attention devoted to them. It is unlikely that resources will quickly be made available to have the organization grow accordingly. So other work probably gets left undone. And there is a multiplier effect here. When noticing that a colleague suffers such consequences for having been involved in a failure, professionals typically start being more cautious with what they document. The paper trails of their actions get larger, more preemptive, more cautious. It is one of the defensive measures that professionals often take. And, as research has shown, paying a great deal of attention to the possibility of being held accountable like this detracts attention and cognitive resources from the actual task.[27] In other words, social workers may be looking harder at paperwork and protocol and procedure than at children.

With all those consequences, is charging the replacement social worker the right thing to do? Consequentialism would suggest not. The things that get changed when a failure is met with an "unjust" response (the prosecution of an individual caregiver in the preceding example) are not typically the things that make the organization safer. They do not typically lead to improvements in primary processes. They can lead to "improvement" of all the stuff that swirls around those primary processes: bureaucracy, involvement of the organization's legal department, bookkeeping, micro-management. Paradoxically, many such measures can make the work of those at the sharp end, those whose main concern is the primary process, more difficult, lower in quality, more cumbersome, and perhaps even less safe.

Golden Rule Ethics

When you are still stumped for what is the right thing to do, think about the golden rule. What that boils down to is this: don't do anything to other people that you would not like to be done to you. Or, put in reverse: the right thing to do is what you would want done unto you. Perhaps that can be your ultimate touchstone. It is also known as common sense ethics. The advantages of this kind of ethical thinking are that it is easy to understand and apply. It also motivates people to do the right thing, because it is a desire they already have (as it applies to them).

Most cultures and ethical or religious traditions have evolved their own version of golden rule ethics, and the principle has been around already for a very long time. Here are some of the versions.

- Plato: *May I do to others as I would that they should do to me.*
- Confucius: *What I do not wish others to do to me, I also do not wish to do to others.*
- Hinduism: *This is the sum of duty: do not do to others what would cause pain if done to you.*
- Judaism: *What you want other people to do to you, do so to them.*
- Christianity: *Do to others as you would have them do unto you.*
- Islam: *You are a believer if you love for the other what you love for yourself.*

Of course, the assumption in golden rule ethics is that what is good for you is also good for the other person. That may not always be true: people may have different values or expectations. But for generic notions of doing good and preventing harm, it may still work very well.

NOT BAD PRACTICE, BUT BAD RELATIONSHIPS

Unjust responses to incidents are less likely the result of bad judgment calls by those who handle the aftermath of an incident and more likely the result of bad relationships between those involved in that aftermath. You can see this in almost any situation where you want to talk of just culture.

Managing relationships between patients and doctors, if not restoring them, is one major aim of mediation, a form of alternative dispute resolution (ADR) in medicine. Mediation restores communication between the two parties, often (if not always) with the help of a mediator. What is said is kept confidential by law, thus making mediation a safe place for showing remorse, for introspection and the exploration of corrective actions without it being seen as admitting liability. In what is called interest-based mediation, in contrast to litigation or criminal-legal action, mediation allows apology, expressions of regret, compassion to occur much more naturally. Mediation is also much more flexible in allowing different outcomes. Compensation does not have to be money (indeed, it most often is not in ADR). In addition to agreeing to care for the injured party in whatever way necessary (medical or otherwise), mediation can inspire changes to procedures, augmenting of education, or other changes that respond to a patient's desire to never see this happen again.[60]

Here is another example of the importance of relationships. Whether employees will see management responses to failure as just depends not so much on the response (or on the bad performance that triggers it). Rather, it depends to a great extent on the existing relationship between management and employees.

We did extensive field work among firefighters to see how they learn from failures that occur during their emergency responses. If firefighters felt that they could come forward with their errors, then it was largely due to the relationship with their supervisors and their managers. It had very little, if anything, to do with formally established procedures or protocol for handling incidents. In one station, firefighters worked in close concert with their management, which had created an atmosphere where reporting errors and suggesting changes was normal, expected, and without jeopardy for any of the parties. Conversely, at a larger urban fire station with distrustful industrial relations, there was less bottom-up participation in decisions involving work context, less firefighter involvement in learning, and much greater suspicion that any reported errors would not be treated fairly.

If bad relationships are behind unjust responses to failure, then good relationships should be seen as a major step toward a just culture. Good relationships are about openness and honesty, but also about responsibility for each other and accountability to each other. Good relationships are about communication, about being clear about expectations and duties, and about learning from each other. Perhaps this can come as somewhat of a relief. "Justice" and "culture" are two huge concepts. They are both essentially contested categories: what either means will forever be open to debate and controversy. They are basically intractable, unmanageable. A relationship, on the other hand, is perhaps more manageable. At least half of it is in your hands. So if you want to do something about just culture, that is where to start.

CASE STUDY

There Is Never One "True" Story

Recall from the preface how "justice," "accountability," and "trust" are essentially contested categories—about which even reasonable people might still disagree after lots of conversation. So we have to acknowledge the existence of multiple ways of thinking about those terms. Because our point of view is not necessarily right, just like nobody else's is. In this final case study, the argument is made that we need to do this for the very notion of "truth" as well. What is *the* "true" account of a failure, of an incident? Is there even such a thing? Previous case studies in this book, like the one about nurse Mara, or even the captain of November Oscar, seem to suggest there wouldn't be. And if there is no one or true account, then relative to which account are we holding people accountable?

Perhaps we should give up trying to dig out the "true" account of a failure altogether. As soon as you make such a claim, somebody will come around and point to "untrue" elements in your story. Or missing parts. Or misconstrued parts, or mischaracterized ones, or underemphasized parts. Trying to tell a story from the perspective "from nowhere" is impossible. As soon as anybody starts describing what happened and what went right or wrong in that story, that person is already using his or her own language, thereby inevitably importing his or her own values, interests, background, culture, traditions, judgments. The courts may have laid a claim on an objective account of a professional's actions. But from the professional's perspective (and that of almost all their colleagues) that account was incomplete, unfair, biased, partial. Remember, in trying to build a just culture, what matters is not getting to a true or objective account of what happened. That is not where the criterion for success lies.

Consider the following story, from the first book of what today is the Hebrew bible (Genesis 19). In this story, a man called Lot is visited by two messengers. They tell him that the city in which he lives, Sodom, is going to be destroyed. "Run, run for your life! Don't look back, don't stop anywhere on the plain. Run to the mountains, lest you be swept away...!" Lot grabs his two daughters and his wife and makes a beeline for the countryside. His two sons-in-law and other family members all dawdle and do not make it out in time.

Then disaster strikes. Fire and brimstone rains down on the city of Sodom and on Gomorrah. The cities are overturned and all their inhabitants are burnt, even the vegetation. Lot's wife looks back, and instantly becomes a pillar of salt.

Lot and his two daughters keep moving. They go up to the mountains outside the town of Zoar, because they fear staying in Zoar. So instead they camp in a cave, Lot and his two daughters.

There, the older daughter says to the younger, "Our father is old and there is no man in the land to come into us in the way of all the world. Let us get our father drunk with wine and sleep with him, that we may quicken with our father's seed."

So they get their father drunk with wine that night and the elder comes and lies with her father. He doesn't know of her lying down or her getting up. The next day the older tells the younger, "Last night I slept with my father. Let us get our father drunk with wine tonight too. You will come and lie with him so you may quicken with our father's seed."

So they get their father drunk with wine that night, too, and the younger got up and lay with him. He doesn't know of her lying down or her getting up. The two daughters of Lot become pregnant from their father. The elder bears a son, whom she calls Moav. He is the ancestor of the Moabites to this day. The younger also bears a son and calls him Ben Ammi. He is the father of the Ammonites of this day.

Suppose the author of this piece of Genesis had told us the story in another way. Suppose the story had told us how Lot, desperate and lonely in the cave, locked up with two helpless and freshly widowed putative virgins. How he forced himself onto his daughters several nights in a row. We may have ended up with acts that were so bad, so morally reprehensible as to amount to "real" crimes. We would have had statutory rape. We would have had incest.

You could even argue that such a story could have been more believable. I mean, how plausible is the story as told in Genesis? Here's an old man, who would have been more than a bit preoccupied with recent bad experiences—losing his wife, his house, his town. Enough to keep an old man awake and a bit distracted. But he seems to sleep just fine. In fact, he sleeps so deeply, as though he's lost to the world. Supposedly made so drunk that he doesn't remember anything of the night. Yet his daughters are capable of tricking him into sexual performances to the point of impregnating them? Twice in a row? While he is essentially unconscious?

But we don't read that Lot raped his daughters. We don't read that Lot committed incest. Instead we read of a plot hatched by the elder daughter, a plot that, we might believe, has a morally justifiable goal: preservation of family lineage after a calamitous interruption that took the men folk out of the ranks and left the remaining women with few prospects (though one might think the town of Zoar wasn't too far). This was perhaps not so strange in a society where family name and tribal affiliation were central to asset ownership and even survival, and where childlessness was a stigmatizing burden.

Whether a crime was committed, then, depends on how the story is told. Even more, it depends on who gets to tell the story. Sutherland's work on white-collar crime, 60 years ago, pointed out that white-collar "crimes" often go under more innocent names such as "fraud." This is possible because they typically violate financial or economic or scientific norms, and inflict no direct bodily harm. There is a difference in punishment too, with fraud generally carrying much lighter sentences than the acts the same society decides to call "crimes." The question here too is, Who gets to say what is what? Sutherland suggested that the difference in naming as well as the difference in sanction afterwards are a consequence of the differential power in society of the populations who commit the different kinds of crime.

This might go for Lot too. Lot was a man. At the time, men generally had more power, more say. And indeed, you may assume that Lot's story was told by a man, not a woman, certainly not one who was young, deeply traumatized, and recently widowed (because then we would likely have heard quite a different story). So what we need to ask is this: If we get somebody to tell the story, and his or her story gets to be the canonical account of the event, whose voice or voices are we not hearing? Who gets sidelined, marginalized, repressed, ignored? Whose view, whose experience is not represented?

A patient died in an Argentine hospital after the use of an experimental US drug, administered to him and a number of fellow patients. The event was part of a clinical trial of a yet unapproved medicine eventually destined for the North American market. To many, the case was only the latest emblem of a disparity where Western nations use poorer, less scrupulous, relatively lightly regulated medical testing grounds in the Second and Third World. But the drug manufacturer was quick to stress that "the case was an aberration" and emphasized how the "supervisory and quality assurance systems all worked effectively." The system, in other words, was safe—it simply needed to be cleansed of its bad apples. The hospital fired the doctors involved and prosecutors were sent after them with murder charges. Calling something murder, in this case, was a choice too. A choice of those with vested interests to protect. A choice of those who had the power to impose their version of events on other people, the version favorable to them and their goals.

Again, whose voices are we not hearing? What other possible perspectives or stories about these deaths are being brushed under the carpet? What are we not learning by labeling these cases as "murder"?

WHICH PERSPECTIVE DO WE TAKE?

If you go to the Louvre in Paris, you may find a painting called *La Méduse*, or *The Raft of the Medusa*, by Theodore Géricault. It's hard to miss, because it's big. Its size, some five by seven meters, is almost overwhelming, sucking you into the ugly real-life scene that depicts a serious accident. An accident report—in a painting. Here is what it shows. Perched on a raft, piled one on top of another and on the remnants of some supplies, are 15 people, some sitting, some sprawling, scattered about the beams. Some are denuded, some in tattered clothing. The raft is adrift on a rough sea, a foamy and angry deep green; waves are rolling in, rolling over the raft. And far, far away, on the horizon there may be a glimpse of something—a rescue ship maybe? One of the survivors has clambered on top of the rubble and on other people and, his arm outstretched, is waving a rag, frantically it seems. The painting shows the man from behind, as if we too are to be hopeful and desperate at the same time, peeking across the raft toward a possible rescue. Who are these people, so emaciated, so close to death? You look at this scene, at the desperation, the destruction, and you can't help but wonder: How did this happen? And you probably ask: Who is to blame, whose fault was this?

The raft is a remnant of the French Navy frigate *La Méduse*. On June 17, 1816, *La Méduse* sailed in a small convoy from Rochefort, headed for St. Louis in Senegal, which the French were going to formally take over from the British. It was carrying 400 people, of whom more than 200 were passengers, including the newly appointed French governor Schmaltz of Senegal. Schmaltz wanted to reach Senegal as quickly as possible. You wonder about his motives, but the man may simply have been eager to claim his new dominion and prevent the British from having second thoughts. Taking the direct route to Senegal would mean skirting the coast of Africa closely, a coast where, in various places, land and sea endlessly merge into one another, with many sandbars and reefs. *La Méduse* was the fastest ship of the convoy and she quickly lost sight of the trailing vessels that were opting for routes farther out to sea.

Commanded by Captain de Chaumareys, and aided by an impromptu navigator (a philosopher named Richefort, who was a member of the Philantropic Society of Cape Verde, an organization dedicated to exploring the African interior), *La Méduse* sailed toward ever shallower water, heralded by white breakers and mud in the water. A lieutenant took it on himself to take soundings off the bow, and, measuring only 18 fathoms (about 30 meters), he warned his captain. Realizing the danger, de Chaumareys ordered his ship turned into the wind, but it was too late. *La Méduse* ran aground on the Bank of Arguin, 50 kilometers off the coast of Mauritania. The accident happened during spring high tide, leaving few possibilities to float the ship again as each subsequent high tide would be lower than the previous one. De Chaumareys also refused to dump its 14 cannons overboard (each weighing three tons).

Plans were made to take everybody to shore with the ship's launches, which would have taken two trips, stretched over a number of days. An alternative idea quickly took shape. It was to build a raft on which the ship's cargo would be offloaded so that *La Méduse* could be floated again. The raft was constructed from masts and crossbeams, measuring 20 meters in length and 7 meters in width and nicknamed *La Machine* by the crew.

But on July 5, a gale developed that threatened to breaking up *La Méduse*. De Chaumareys decided to immediately evacuate the frigate, with 146 men and one woman boarding the woefully inadequate raft, which was to be towed ashore by the lifeboats of *La Méduse*. The raft had few supplies and no method of navigation or steering. Much of its deck was under water. Seventeen men decided to stay on *La Méduse*. The rest boarded the lifeboats. They quickly realized that towing the raft was a hopeless endeavor, and began to fear that they would be overwhelmed by its desperate survivors. The raft was cut loose, leaving the 147 occupants adrift. Those in the lifeboats made it safely to the coast of Africa, with most finding their way overland to Senegal, though some died on the way.

Conditions on the raft quickly became wretched. Among the few provisions were casks of wine instead of water. Different factions developed, with officers and passengers in one, soldiers and sailors in another. Fights broke out. On the first night, 20 men lost their lives either through suicide or murder. Stormy weather kept threatening and people continually scrambled to get to the center of the raft. Many were pushed or washed overboard in the scuffles.

After four days, fewer than half the survivors were still on board. Rations, such as they were, were dwindling rapidly, and some people resorted to cannibalism. On the eight day, the fittest began throwing the weaker and wounded overboard until only 15 remained. On July 17, they sighted one of the ships in the original convoy. It disappeared over the horizon, plunging the survivors into profound gloom. The ship reappeared two hours later, however, and they were finally rescued after 13 days adrift. Five of the survivors died within days.

Henri Savigny, the surviving surgeon of *La Méduse*, submitted his account to the French authorities only a few months after the shipwreck. It was leaked to an anti-Bourbon newspaper, *Le Journal des Débats*. Together with Alexander Corréard, a geographer, he then wrote a book about the case. It became a hit, going through five editions within the next years, and was translated into English, German, Dutch, and Italian. With every edition, the message of the book became more political.

A scandal became unavoidable, one heavily drenched in French politics. In 1815, Napoleon had been defeated at Waterloo and exiled, allowing the restoration of the Bourbon monarchy that had been terminated by the French Revolution of 1792. Its restoration brought the imperative to reward loyalists and populate key positions with confidantes. Viscount Hugues Duroy de Chaumareys was one of them and, after requesting a naval post from the king's brother, he was given the command of *La Méduse*. At 53, he had hardly been on a ship in the preceding 25 years, and had never commanded one, instead working as a customs officer.

In 1817, de Chaumareys was court-martialed. He was found not guilty of abandoning his ship, nor of failing to refloat his ship, nor of abandoning the raft. He was, however, convicted because of incompetent and complacent navigation and of abandoning *La Méduse* before all passengers had been taken off. The verdict could have led to the death penalty, but he was given three years in prison—a whitewash according to some. A year later, Governor Schmaltz of Senegal was forced to resign. The Gouvion de Saint-Cyrl law later ensured that appointments and promotions in the French military were more merit-based than before.

Inspired by stories of the shipwreck and the subsequent scandal, Theodore Géricault, a 25-year-old artist, decided to create his own depiction of it. His painting shows the moment that was recounted to him by one of the survivors: just as the rescue ship was disappearing over the horizon. Whatever had not yet been told in *Le Journal*, the Savigny-Corréard book, or any of the other stories swirling around the shipwreck, was now on full display in picture-perfect form: a monumental disaster with a bad aftermath, a painting that in itself formed an indictment against a corrupt, insider monarchist establishment. Géricault's painting tells one story. It is for sure a valid one: the desperation and anxiety on the part of the few survivors must have been all too real for them. But again we need to ask: Whose version of the events is not seen here?

Suppose Géricault had created a painting in which brave officers and passengers are valiantly trying to tow the raft, but in which the raft's occupants are looming over them, depicted as numerous and livid, as sinister and menacing, threatening to overwhelm the launches and lynch or scuttle anybody still on them. Such a painting could tell the story from the point of view of the occupants of the launches. It may well have inspired sympathy with them and their plight, and perhaps even supported public understanding for their decision to cut the raft loose.

Similarly, a painting that would have given prominence to the gathering storm that prompted De Chaumereys to order the abandonment of *La Méduse* could have helped explain that decision and made it look less culpable. The culpability arises in part because of hindsight: De Chaumereys later went back to the shipwreck to try to recover the gold that was believed to be still on board (as well as a few survivors who had decided to stay behind). To his probable surprise, *La Méduse* was still intact, despite the storm. In hindsight, the decision to abandon ship was unwise, perilous, premature.

A painting that shows the actual moment of that decision, with a really bad storm about to barrel down on the stuck ship, could have helped set it in context and make De Chaumereys look much better. Details about his life, in which his monarchist affiliation had made it impossible for him to attain a naval post and in which not having sailed in 25 years was not a condition of his own making, could gain greater notability there too. It wasn't that he was uniquely unqualified for the job; rather, he had been denied the possibility to exercise it for a while because of contemporary political configurations.

Again, whether a crime was committed depends on how the story is told, and on who gets to tell the story. De Chaumerays may have gotten a tough treatment in Géricault's depiction of the aftermath of his decisions. Those in power, or legitimated by society, to decide whether his behavior amounted to punishable crimes, however, came to a different conclusion. In their story of the events, he was found not guilty of the most incriminating counts, and punished mildly for the remaining lesser accusations. De Chaumerays' royalist dedication would probably have not hurt here in a country once again under the reign of a Bourbon. Was a crime committed, and if so, which acts constitute enough badness to amount to a crime? Well—who gets to say? Who gets to tell the story?

The "Real" Story of What Happened?

Géricault certainly tried to tell the "real" story of what happened. His portrait of *La Méduse* was an important installment in the development of the Realist school of art, a school concerned with depicting accurately and objectively the scene or event. Realism departed from Romanticism in deliberately rejecting subjects that were made to look more beautiful, wistful almost, than they could ever be in real life (like the king regally sitting on his horse in full ornamentation). Realism also dealt with subjects deemed inappropriate to nineteenth-century sensibilities (the nudity in Géricault's painting would have added to its shock value). Realism took off in the nineteenth century in France, with, for example, Gustave Courbet. It tried to capture the "real," the sincere, the ugly, unsentimental, and unidealized version of contemporary life.

The aim may be sincere, and a welcome departure from the deliberately dressed-up, beautified, embellished versions of a subject of painting or writing. But with the commitment to show how things "really" were comes epistemological arrogance. Who is to say that this is "really" how it was? How do you know? And if one depiction is supposed to be the real one, are we to conclude that all other depictions are false? One person shows the truth; all others are liars?

Realist paintings such as Géricault's succeed only in capturing one moment, from one angle. An angle on a scene reveals something, for sure, but it hides even more. Each perspective discloses and obscures at the same time. Somebody paints an accident scene at one moment, from one perspective, and all other perspectives and moments are lost. You don't know about them, you don't hear them, you don't see them.

Early twentieth-century schools of art such as cubism, in a reaction to this, tried to capture multiple perspectives simultaneously. They used geometric shapes, collage, or interlocking planes, Picasso and Cézanne being the obvious exponents. Cubism rejected the notion that we can represent reality only from one single viewpoint at a time. It challenged the assertion by Realists that they could "accurately" and "objectively" depict an event or a scene. Such supposed accuracy and objectivity again implies epistemological hubris, as if some people have privileged access to the "real" account, the objectively "true" one, and that they are therefore entitled to impose that on other people too.

RHETORIC: THE ART OF PERSUASION

La Méduse is a persuasive painting. The images in it seem to say: Look, here is what happened; here is the unvarnished, ugly truth. It is rhetoric, something we normally associate with words, but in this case it is rendered in brushstrokes. In the West, we have all learned that factual evidence, expressed plainly and clearly (as was indeed the ambition of the Realist school, and is in much of our scientific and judicial traditions) is the most convincing form of knowledge. Much more so than pretty arguments, or nice stories, or emotional appeals, or psychological trickery. Facts speak for themselves; just look at *La Méduse*. See what happened? See how bad this is, how morally reprehensible? How *criminal*?

This view assumes that a phenomenon such as a crime is real, that an act that is factually criminal is inherently criminal. From this view, crime is more than simply telling or painting the story one way and not another. Crime represents an identifiable form of behavior caused by an identifiable set of mechanisms and factors (there may be violence, the uncontrolled release of energy, other people getting hurt as a result). Crime is not the contingency of who gets to tell about what happened. Crime has unavoidable material features that cannot be defined away. In other words, the act has the immutable identity of a crime. It doesn't matter who commits it, or where or when, or under what duress, and it certainly doesn't matter who gets to tell the story about the act. A crime is a crime, no matter how you describe it, or who gets to describe it. And, while at it, a criminal is a criminal. He or she doesn't get made into one by the descriptions of other people around him or her.

From this position, Lot was a serial rapist. From this position, De Chaumereys was an incompetent, unqualified moron. He was that before the trip of *La Méduse*, during the trip, after the trip. He weaseled his way into the job, consistent with who he was. And during the trip, he predictably engaged in something that was criminally negligent, also consistent with who he was. Turning him into a criminal is not the result of a label that other people put on him. It is something he himself did, or even was. Similarly, from this perspective, the pilots in Chapter 4 were unprofessional, and really did engage in willful violations. The pharmacist really did commit something criminal when he didn't double-check the lethal solution prepared by his lab tech.

But the compelling question, raised by Lot and the Argentine doctors and *La Méduse* is still: Who gets to tell the story? And how and when does its content and tone amount to a story of a punishable crime? Remember, Géricault might have chosen to paint another angle, another moment of the whole event, and the supposedly criminally negligent aspects of De Chaumereys or his actions may have dissipated. And who got to describe the act so that it became a crime (Géricault; the drug company doing its trials in Argentina)? Or, for that matter, who got to describe the act so that it didn't amount to a crime (Lot)? The focus from this position is not on who committed the so-called crime, but on who called it so.

Central to the creation of such stories is the lack of status or power on the part of those who are set to lose; those who do not have a voice are on the losing end of the construction. Lot's daughters lacked status and power, and did not have a voice (or were denied one in Genesis), so their father did not commit a crime. The drug company had status and power, so the doctors administering its trials in Argentina, who didn't have much of a voice in this debacle, committed a crime. De Chaumereys (or his backers) still had enough power and status and voice: his "crime" was judged very lightly. Dead pilots lack power or status and have no voice, so it is easy to convert their acts into bad, sanctionable behavior posthumously. Mara still had a voice, but by the time it was heard in court, it was rendered feeble and partisan. It came from a prepared transcript, squeezed into the tightly scripted and formal proceedings of a court case, where the words it uttered became easily seen as an evasive, self-serving, and exculpatory maneuver.

This means that justice and power are closely overlapping categories. We cannot consider one without the other.

References

1. Sharpe VA. Promoting patient safety: An ethical basis for policy deliberation. *Hastings Center Report* 2003;**33**(5):S2–19.
2. Reason JT. Are we casting the net too widely in our search for the factors contributing to errors and accidents? In: Misumi J, Wilpert B, Miller R, eds. *Nuclear safety: A human factors perspective*. London: Taylor & Francis, 1999;210–23.
3. Wachter RM, Pronovost PJ. Balancing "no blame" with accountability in patient safety. *New England Journal of Medicine* 2009;**361**:1401–06.
4. Dekker SWA, Hugh TB. Balancing "no blame" with accountability in patient safety. *New England Journal of Medicine* 2010;**362**(3):275.
5. Young MS, Shorrock ST, Faulkner JPE et al. Who moved my (Swiss) cheese? The (r)evolution of human factors in transport safety investigation. Gold Coast, QLD, Australia: International Society of Air Safety Investigators, 2004.
6. Levitt P. When medical errors kill: American hospitals have embraced a systems solution that doesn't solve the problem. March 19, 2014. Retrieved from: http://www.latimes.com/opinion/commentary/la-oe-levitt-doctors-hospital-errors-20140316,0,4542704.story.
7. Shojania KG, Dixon-Woods M. 'Bad apples': Time to redefine as a type of systems problem? *BMJ Quality and Safety* 2013;**22**(7):528–31.
8. Dekker SWA, Leveson NG. The systems approach to medicine: Controversy and misconceptions. *BMJ Quality and Safety* 2015;7–9.
9. Dekker SWA. *Second victim: Error, guilt, trauma and resilience*. Boca Raton, FL: CRC Press, 2013.
10. Newland J. Medical errors snare more than one victim. *The Nurse Practitioner* 2011;**36**(9):5.
11. Cohen-Charash Y, Spector PE. The role of justice in organizations: A meta-analysis. *Organizational Behavior and Human Decision Processes* 2001;**86**(2):278–321.
12. Conquitt JA, Conlon DE, Wesson MJ et al. Justice at the millennium: A meta-analytic review of 25 years of organizational justice research. *Journal of Applied Psychology* 2001;**86**:425–45.
13. GAIN. *Roadmap to a just culture: Enhancing the safety environment*. Global Aviation Information Network (Group E: Flight Ops/ATC Ops Safety Information Sharing Working Group), 2004.
14. Seed G, Palmer A. A shot in the dark. *The Sunday Telegraph* January 24, 1999;23.
15. Wilkinson S. The November Oscar incident. *Air & Space* 1994;80–87.
16. Rawls J. *A theory of justice*. Cambridge, MA: Harvard University Press, 2003.
17. Berlinger N. *After harm: Medical error and the ethics of forgiveness*. Baltimore, MD: Johns Hopkins University Press, 2005.
18. Eurocontrol. Report on legal and cultural issues in relation to ATM safety occurrence reporting in Europe: Outcome of a survey conducted by the Performance Review Unit in 2005–2006. Brussels: Eurocontrol Performance Review Commission, 2006.
19. Ferguson J, Fakelmann R. The culture factor. *Frontiers of Health Services Management* 2005;**22**(1):33–40.
20. Fischhoff B. Hindsight ≠ foresight: The effect of outcome knowledge on judgment under uncertainty. *Journal of Experimental Psychology: Human Perception and Performance* 1975;**1**(3):288–99.

21. von Thaden T, Hoppes M, Yongjuan L et al. The perception of just culture across disciplines in healthcare. In: Society HFaE, ed. *Human Factors and Ergonomics Society 50th Annual Meeting*. San Francisco: Human Factors and Ergonomics Society, 2006;964–68.
22. Frederick J, Lessin N. The rise of behavioural-based safety programmes. *Multinational Monitor* 2000;**21**:11–17.
23. Rowe M. People who feel harassed need a complaint system with both formal and informal options. *Negotiation Journal* 1990;**6**(3):161–72.
24. Denham CR. TRUST: The 5 rights of the second victim. *Journal of Patient Safety* 2007;**3**(2):107–19.
25. Wahlberg D. Report: Systemic problems at St. Mary's set stage for nurse's fatal drug error. *Wisconsin State Journal* March 15, 2010;8.
26. *New York Times*. Breaking the rules to get more recruits: Some say cheating needed to fill ranks. New York, p. 4 (US National), May 4, 2005.
27. Lerner JS, Tetlock PE. Accounting for the effects of accountability. *Psychological Bulletin* 1999;**125**(2):255–75.
28. Ashford JB, Kupferberg M. *Death penalty mitigation: A handbook for mitigation specialists, investigators, social scientists, and lawyers*. Oxford: Oxford University Press, 2014.
29. Dekker SWA. *The field guide to understanding 'human error'*. Farnham, UK: Ashgate, 2014.
30. Dekker SWA, Laursen T. From punitive action to confidential reporting: A longitudinal study of organizational learning. *Patient Safety & Quality Healthcare* 2007;**5**:50–56.
31. Braithwaite J. *Crime, shame and reintegration*. Cambridge, UK: Cambridge University Press, 1989.
32. Sharpe VA. *Accountability: Patient safety and policy reform*. Washington, DC: Georgetown University Press, 2004.
33. Delbanco T, Bell SK. Guilty, afraid, and alone: Struggling with medical error. *New England Journal of Medicine* 2007;**357**(17):1682–83.
34. Thomas G. Aviation on trial. *Air Transport World* 2002;31–33.
35. Wu AW. Medical error: The second victim. *BMJ* 2000;**320**(7237):726–28.
36. Hilfiker D. Facing our mistakes. *New England Journal of Medicine* 1984;**310**:118–22.
37. Leonhardt J, Vogt J. *Critical incident stress management in aviation*. Aldershot, UK: Ashgate, 2006.
38. Friel A, White T, Alistair H. Posttraumatic stress disorder and criminal responsibility. *Journal of Forensic Psychiatry and Psychology* 2008;**19**(1):64–85.
39. Serembus JF, Wolf ZR, Youngblood N. Consequences of fatal medication errors for healthcare providers: A secondary analysis study. *MedSurg Nursing* 2001;**10**(4):193–201.
40. Berlinger N. Avoiding cheap grace: Medical harm, patient safety and the culture(s) of forgiveness. *Hastings Center Report* 2003(November–December):28–36.
41. Buikema M. *This never again (Dit nooit meer)*. Utrecht, NL: CBO, Kwaliteitsinstituut voor de Gezondheidszorg, 2010.
42. Ring DC, Herndon JH, Meyer GS. Case 34-2010: A 65-year-old woman with an incorrect operation on the left hand. *New England Journal of Medicine* 2010;**363**:1950–57.
43. Zehr H, Gohar A. *The little book of restorative justice*. Intercourse, PA: Good Books, 2002.
44. Bosk C. *Forgive and remember: Managing medical failure*. Chicago: University of Chicago Press, 2003.
45. Christie N. *A suitable amount of crime*. London: Routledge, 2004.
46. Becker HS. *Outsiders: Studies in the sociology of deviance*. London: Free Press of Glencoe, 1963.

47. *PR Newswire*. LIJ Medical Center Introduces One-of-a-Kind Video Monitoring Project to Enhance Patient Safety in OR's. New Hyde Park, NY: PRNewswire-USNewswire, pp. 1–2, May 16, 2014.
48. Starbuck WH, Milliken FJ. Challenger: Fine-tuning the odds until something breaks. *The Journal of Management Studies* 1988;**25**(4):319–41.
49. McDonald N, Corrigan S, Ward M. Well-intentioned people in dysfunctional systems. Fifth workshop on human error, safety and systems development. Newcastle, Australia, 2002.
50. Burnham JC. *Accident prone: A history of technology, psychology and misfits of the machine age*. Chicago: University of Chicago Press, 2009.
51. Prokscha S, Strzepek A. *Question marks over EU register of bad doctors*. EUObserver. May 25, 2016. Retrieved from: http://www.euobserver.com/social/119244, 2013.
52. Dekker SWA. *Patient safety: A human factors approach*. Boca Raton, FL: CRC Press, 2011.
53. Saines M, Strickland M, Pieroni M et al. Get out of your own way: Unleashing productivity. In: Vorster G, Richardson C, Redhill D, eds. *Building the lucky country: Business imperatives for a prosperous Australia*. Sydney, Australia: Deloitte Touche Tohmatsu, 2014.
54. Weick KE, Sutcliffe, KM. *Managing the unexpected: Assuring high performance in an age of complexity*. San Francisco: Jossey-Bass, 2001;118.
55. Grotan TO. Hunting high and low for resilience: Sensitization from the contextual shadows of compliance. In: Steenbergen RDJM, van Gelder PHAJM, Miraglia S et al., eds. *Safety, reliability and risk analysis: Beyond the horizon*. Boca Raton, FL: CRC Press, 2014;327–34.
56. Hollnagel E. *Safety I and Safety II: The past and future of safety management*. Farnham, UK: Ashgate, 2014.
57. Snook SA. *Friendly fire: The accidental shootdown of US Black Hawks over northern Iraq*. Princeton, NJ: Princeton University Press, 2000.
58. Solomon RC, Flores F. *Building trust: In business, politics, relationships and life*. Oxford: Oxford University Press, 2003.
59. Brennan TA, Sox CA, Burstin HR. Relation between negligent adverse events and the outcomes of medical malpractice litigation. *New England Journal of Medicine* 1996;**335**:1963–67.
60. Dauer EA. Ethical misfits: Mediation and medical malpractice litigation. In: Sharpe VA, ed. *Accountability: Patient safety and policy reform*. Washington, DC: Georgetown University Press, 2004;185–202.
61. Hidden A. Clapham Junction accident investigation report. London: Her Majecty's Stationery Office, 1989.
62. Lanir Z. The reasonable choice of disaster: The shooting down of the Libyan airliner on 21 February 1973. *Journal of Strategic Studies* 1989;**12**:479–93.
63. Barach P, Small SD. Reporting and preventing medical mishaps: Lessons from non-medical near miss reporting systems. *BMJ* 2000;**320**(7237):759–63.
64. Vaughan D. *The Challenger launch decision: Risky technology, culture, and deviance at NASA*. Chicago: University of Chicago Press, 1996.
65. Billings CE. *Aviation automation: The search for a human-centered approach*. Mahwah, NJ: Lawrence Erlbaum Associates, 1997.
66. Friendly fire death was a criminal act, coroner rules. *The Guardian* March 18, 2007;1.
67. Goglia J. Southwest Airlines settles whistleblower suit by mechanic disciplined for reporting cracks in 737. *Forbes* February 11, 2015;6.
68. NSWHealth. Open disclosure guidelines. Sydney, NSW: Department of Health, 2007;29.
69. HogstaDomstolen. Verdict B 2328-05. In: Court HDSS, ed. Stockholm: Swedish Supreme Court, 2006;1–6.

70. Dekker SWA. The criminalization of human error in aviation and healthcare: A review. *Safety Science* 2011;**49**(2):121–27.
71. Govan F. Mechanics face manslaughter charge for Madrid air crash. *The Telegraph* October 16, 2008;12.
72. Wagenaar WA. *Vincent plast op de grond: Nachtmerries in het Nederlands recht* (Vincent urinates on the ground: Nightmares in Dutch law). Amsterdam: Uitgeverij Bert Bakker, 2006.
73. Rowe M. The rest is silence. *Health Affairs* 2002;**21**(4):232–36.
74. *New York Times.* 4 convicted in 2001 Milan plane crash. March 15, 2005;5.
75. NTSB. Accident report: Crash of Pinnacle Airlines Flight 3701, Bombardier CL600-2-B19, N8396A, Jefferson City, Missouri, October 14, 2004. Washington, DC: National Transportation Safety Board, 2007.
76. McKenna JT. Criminal, safety probes at odds. *Aviation Week & Space Technology* 1999;**151**(24):47–48.
77. Ognibene S, Binnie I. Costa Concordia captain sentenced to 16 years for 2012 shipwreck. *Reuters* February 11, 2015. Retrieved from: http://www.reuters.com/article/us-italy-ship-idUSKBN0LF12H20150211 (accessed on February 20, 2015).
78. Kaminski-Morrow, D. Italian safety agency accuses judicial authorities of hampering investigation into Cessna executive jet accident. Air Transport Intelligence News, p. 1. London: Flight Global, February 26, 2009.
79. North DM. Oil and water, cats and dogs. *Aviation Week & Space Technology* 2002;**156**(5):70.
80. Ruitenberg B. Court case against Dutch controllers. *The Controller* 2002;**41**(4):22–5.
81. Ter Kulle A. Safety versus justice. *Canso News* 2004;**18**:1–2.
82. Bittle S, Snider L. From manslaughter to preventable accident: Shaping corporate criminal liability. *Law & Policy* 2006;**28**(4):470–96.
83. Merry AF, McCall Smith A. *Errors, medicine and the law.* Cambridge, UK: Cambridge University Press, 2001.
84. Studdert DM, Thomas EJ, Burstin HR et al. Negligent care and malpractice claiming behavior in Utah and Colorado. *Medical Care* 2000;**38**(3):250–60.
85. Thomas EJ, Studdert DM, Burstin HR et al. Incidence and types of adverse events and negligent care in Utah and Colorado. *Medical Care* 2000;**38**(3):261–71.
86. Kessler D, McClellan M. Do doctors practice defensive medicine? *The Quarterly Journal of Economics* 1996;**111**(2):353–90.
87. Nagel T. *The view from nowhere.* Oxford: Oxford University Press, 1992.
88. Merry AF, Peck DJ. Anaesthetists, errors in drug administration and the law. *New Zealand Medical Journal* 1995;**108**:185–87.
89. Allard T. Garuda death crash pilot jailed. *The Sydney Morning Herald* April 7, 2009.
90. Schmidt S. Some airlines require confidentiality deals before safety inspections. *Canwest News Service* February 25, 2009;6.
91. Hollinger P. Dutch to publish details of Schiphol air traffic incidents. *Financial Times* January 13, 2015. Retrieved from: http://www.ft.com/cms/s/0/195c797c-9b1d-11e4-b651-00144feabdc0.html#axzz469Kz5a2W (accessed on January 20, 2015).
92. McGee E. *Aviation Safety groups issue joint resolution condemning criminalization of accident investigations.* Washington, DC: Air Safety Foundation. May 25, 2016. Retrieved from: http://flightsafety.org/media-center/press-releases/2006-press-releases/aviation-safety-groups-issue-joint-resolution-condem, 2006.
93. Flight Safety Foundation. Aviation Safety Groups Issue Joint Resolution Condemning Criminalization of Accident Investigations [Press release]. October 18, 2016. Retrieved from: http://flightsafety.org/media-center/press-releases/2006-press-releases/aviation-safety-groups-issue-joint-resolution-condem.

94. Phipps J, Puplidy I. Saddleback fire learning review. Washington, DC: US Forest Service, 2012.
95. Morreim EH. Medical errors: Pinning the blame versus blaming the system. In: Sharpe VA, ed. *Accountability: Patient safety and policy reform*. Washington, DC: Georgetown University Press, 2004;213–32.
96. Allspaw J. Blameless postmortems and a just culture. In: Etsy, ed. *Code as craft*. New York: Etsy, 2012.
97. *International Herald Tribune*. Texas executives faulted in BP explosion. May 4, 2007;10.

Index

G

H

I

J

K

L

S

T

U

V

W

Printed in the United States
by Baker & Taylor Publisher Services